020442

A Foundation Course in Statics and Dynamics

D0178314

A Foundation Course in Statics and Dynamics

DAVID PLUM AND MARTIN DOWNIE

 LONGMAN

© Addison Wesley Longman 1997

Addison Wesley Longman Limited
Edinburgh Gate
Harlow
Essex CM20 2JE
England

and Associated Companies throughout the World.

The rights of David Plum and Martin Downie to be identified as authors of this Work have been asserted by them in accordance with the Copyright, Designs and Patents Act 1988.

All rights reserved. No part of this publication may be reproduced, stored in a retrieval system, or transmitted in any form or by any means, electronic, mechanical, photocopying, recording or otherwise, without either the prior written permission of the publisher or a licence permitting restricted copying in the United Kingdom issued by the Copyright Licensing Agency Ltd, 90 Tottenham Court Road, London W1P 9HE.

Text design by Claire Brodmann
Typeset by 54 in 10/12pt Sabon
Printed and bound by Henry Ling Ltd, at the Dorchester Press, Dorset

First printed 1997

ISBN 0582 21060 7

Library of Congress Cataloging-in-Publication Data

Plum, David.
    A foundation course in statics and dynamics / David Plum and
Martin Downie.
        p.   cm.
    Includes index.
    ISBN 0-582-21060-7 (alk. paper)
    1. Mechanics, Analytic.   I. Downie, Martin.   II. Title.
QA807.P655   1997   97-19518
531—dc21   CIP

A catalogue record for this book is available from the British Library.

Contents

v

Preface

The aim of this book is to bridge the gap, which has steadily widened over the last decade, between the syllabuses for A-level mathematics and the requirements of first-year university courses in engineering and technology. This gap has occurred as a result of three movements:

- A-level syllabuses have reduced the force vector, statics and motion content of the most commonly taken mathematics courses. Much of what might have been described as applied mathematics is now to be found in specialist (second) mathematics courses. In the mid-eighties the JMB reported a drop from 7000 to 4000 candidates sitting this topic.
- In the late eighties a government-sponsored initiative encouraged a national movement to bring candidates into technology courses from a broader A-level background. This resulted in many university entrants to engineering and technology having an inadequate mathematics base. It gave rise to additional introductory courses at university in the form of foundation years, conversion courses, and so on.
- The more recent debate in the mid-nineties has highlighted the paradox between the increasing success of candidates at A-level mathematics and the increasing need for remedial classes at university. It has been suggested that this is most readily explained by the fact that syllabuses have widened and become less deep. Some topics can be dealt with in conceptual terms only, without the need for analytical ability.

Analytical ability and clear concepts are what is required for the innovative demands of university courses in engineering and technology. The aim of this book, therefore, is to enable university entrants who are weak in the applied mathematics topics to be better prepared for their first-year courses. It will be valuable as a text for introductory courses and as a reinforcement to the basic texts in engineering courses which presume prior knowledge of much foundational material.

The chapters are arranged to take the student from basic concepts of force, equilibrium, motion and energy through to their application in framed structures and force systems, rectilinear and curvilinear motion, friction, work and momentum. The basic relationships are described in each chapter and clarified by diagram or equation as appropriate. Topics are illustrated by worked examples and test-yourself problems. Recap boxes reinforce and summarize the principal teaching objectives at key points in each chapter. The text, examples and problems are all set, as far as possible, in the context of real engineering applications, which will help the student to visualize and remember the principles being explained.

The authors trust they have eliminated all the errors from the calculations. Identification of any errors will be gratefully received.

Acknowledgements

The authors would like to acknowledge the debt owed to many teachers and lecturers who over many years taught them the basics of statics and dynamics. The subject matter has a long history, but is as fundamental today as it was in its misty beginnings. Without a sound understanding of statics and dynamics, much of what we now call technology is flawed. What was taught to us, we now pass on, but in the process we have adapted these historical subjects to the needs of today.

Introduction

1.1 Statics and dynamics

Statics and dynamics are subjects which form the basis for understanding the action of forces, the nature of movements, and the way in which these phenomena interrelate with each other. Considerable parts of engineering depend on an accurate appreciation of the nature of forces and movements, so statics and dynamics are fundamental to many disciplines of engineering and technology. They have particular importance to aeronautical engineering, civil engineering, marine technology, mechanical engineering, and structural engineering. These branches of engineering and technology interrelate largely through their common foundations in statics and dynamics.

Statics is primarily concerned with the behaviour of forces acting on objects at rest. It is therefore of prime importance when considering the stability of buildings and similar large structures. But it is of equal significance when considering very small components such as switches and control mechanisms. And statics is just as relevant for components in buildings, aircraft, cars, ships and domestic appliances. Although statics usually assumes a state of rest for the object or component, it has equal meaning for states of uniform motion. And it may be applied to states of varying motion by use of equivalent static forces.

Dynamics is the study of the motions of objects and the forces that cause them. If the forces acting on an object or a system are not in equilibrium (balance each other out), it will be set in motion. The basic rules for predicting the behaviour of unbalanced systems were largely developed by Sir Isaac Newton towards the end of the seventeenth century. The application of these rules, called Newtonian mechanics, is covered in Parts II and III. The subject naturally divides into kinematics and kinetics. Kinematics studies the motions of objects without reference to the forces causing them. Kinetics studies the relationship between the motions of objects and the forces to which they are subjected. Nearly all engineering systems respond dynamically at some level. Sometimes the dynamics of a system are of supreme importance, as in the design of machines; sometimes they can be almost ignored. The twin towers of the World Trade Center in New York are so tall they can be felt swaying in the wind (responding dynamically to wind forces) but they were designed largely upon the principles of statics. Because of its relevance to virtually all aspects of engineering design and application, a thorough understanding of the subject is essential for scientists and engineers.

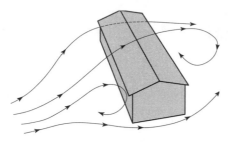

Figure 1.1 The relationship between wind speed, direction and pressure may follow a statistical distribution.

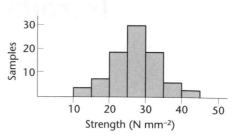

Figure 1.2 Strength of concrete on a construction site.

➤ 1.2 Laws of physics

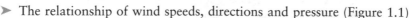

Many natural phenomena take place in accordance with clearly defined laws. They have repeatability; if they were observed on more than one occasion, each time they would demonstrate the same relationship. In all such events there are some factors which are not readily controlled, so perfect repeatability is impossible to achieve. Our observations of natural phenomena therefore always contain an element of variability; here are some examples:

➤ Observation of the acceleration of a falling body
➤ The relationship of extension to applied force for a particular material
➤ Variation of volume, pressure and temperature for a gas

Minor variations in the object being observed, the surrounding influences and imperfections in our methods of observation, all will lead to some degree of uncertainty in the result. Any laws which may be proposed to express the observed events therefore contain this same degree of uncertainty. Physical laws may therefore be described as **best-fit statements** of observed events. Statements of this kind are essential to the development of engineering and technology, and for convenient use are presented in the form of equations or graphs.

Although many observed events appear to accord with clear laws, allowing for these variations, some other events may not show such clear relationships. Observation may show a wide scatter of results, which might be described as a **distribution** or **population**; here are some examples:

Figure 1.3 An irregular body falling in a viscous fluid.

➤ The relationship of wind speeds, directions and pressure (Figure 1.1)
➤ Variation of the strength of concrete on a construction site (Figure 1.2)
➤ Observation of an irregular body falling in a viscous fluid (Figure 1.3)

For these and similar events the distributions of data could be assembled using statistical methods. Laws derived in this way would have much wider degrees of uncertainty, which themselves would need to be stated and quantified. Any physical law contains *some degree* of uncertainty, whether it be to a minor extent or a major extent.

Equations or graphs may be used to express laws governing physical events and may be simple or more complex in form:

- Force = mass × acceleration ($F = ma$)
- Final velocity = initial velocity + acceleration× time ($v = u + at$)
- Elastic modulus = stress/strain ($E = \sigma/\varepsilon$)

Scientists and engineers are interested in describing, recording and communicating physical processes for use in engineering design. If such information has been collected by observation it is usually processed in the form of discrete data. That is to say, measurements of quantities of interest are repeatedly taken and related to some independent variable such as time, displacement or temperature. In this case the data is only available at those instants when the measurements were made. Some experiments, to all intents and purposes, have output in the form of continuous data, where the data is recorded by equipment such as a pen recorder, which presents the information in the form of a continuous curve or signal. In these cases the data is available at any point during the period over which the experiment was conducted. If the data is well behaved it can often be represented by a curve. The curve's equation can be deduced from the data and sometimes a physical law may be formulated on the strength of it.

One set of physical laws fundamental to the understanding of statics and dynamics are Newton's laws of motion. These may be stated as follows:

(1) A particle remains at rest, or continues at uniform velocity in a straight line, unless acted on by a force.
(2) The acceleration of a particle is proportional to the force acting on it and is in the direction of the force.
(3) The forces of action and reaction between particles are equal in magnitude, opposite in direction, and are collinear.

These three laws will be explored and developed in later chapters, particularly Chapters 3 and 7, but the concepts expressed in them form the foundations of this text, and for statics and dynamics in general.

1.3 Concept and calculation

Physical phenomena and the laws describing them are discussed in Section 1.2. They are usually the result of observing natural events, and attempting to formulate an idea which governs their behaviour as seen. This idea, encapsulated in a law, is then passed on to others in the form of words, diagrams, models or even demonstrations. This is the **concept**; it must be capable of transmission in a form that is understandable to its audience. Because concepts embodied in physical laws describe actual natural events, some degree of visualization is possible, and even essential, to understanding the topic. A concept may be expressed in words or equations, but to make certain it is understood, any concept should be converted into visual form.

Calculation by contrast does not require understanding of the concepts contained in a physical law, but merely its application to obtain some desired information. All that is required in the calculation process is an acceptance of some equations, purporting to describe natural events, coupled with an ability in algebra and arithmetic. At the beginning of a course in engineering it is possible to make progress and even pass examinations by using calculation allied to a good memory. But to make secure progress in engineering it is essential to have grasped the concepts fully. In this sense the ability to learn rules of calculation, and to carry out the process successfully, may be a barrier to a proper understanding of the concepts.

However, once the proper concepts are visualized and grasped, calculation of examples will reinforce an understanding. Calculation will also give substance to the concepts, ensure that relative magnitudes are recognized, and give an appreciation of the correct units.

Calculation of specific information from given equations is usually carried out simply by use of a hand calculator. The only requirement for accuracy is to know the functions of the calculator and the proper sequences for inputting data. Most calculators will also include functions to process data in statistical form.

The use of computers in calculation is of greatest value when the equations and arithmetic become complex, or when one calculation process is to be repeated many times. Neither requirement is appropriate to examples in this text, but it is always worthwhile to obtain practice in computing, especially inputting data and presenting results.

➤ 1.4 Accuracy

The variation in most engineering data (Section 1.2) means it is usual for it to carry a limited accuracy. Hence information given for examples will usually be presented to two-figure or at most three-figure accuracy. No amount of calculation can increase that accuracy, and it is therefore inappropriate to give answers to a greater accuracy. Examples of the three-figure accuracy are numbers such as 1.27, 0.0164, 81 600 or 20.2.

Presentation of data in the form of graphs can lead to calculations which are themselves approximate. Where graphs are given in the form of defined mathematical relationships then slopes and areas associated with the graphs can be found precisely, as in Figures 1.4 and 1.5. Where the graph represents a more random set of data, as in Figure 1.6, calculation of slopes and areas may introduce an element of approximation. The accuracy of representing the physical process described by the

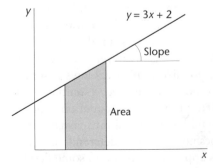

Figure 1.4   Straight line: slopes and areas can be found precisely.

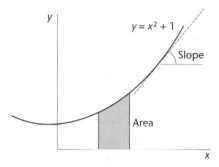

Figure 1.5 Quadratic curve: slopes and areas can be found precisely.

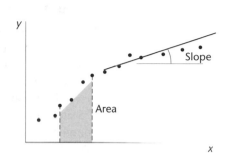

Figure 1.6 Random data: slopes and areas can only be estimated.

variable y depends on how rapidly it changes with x. Chapter 7 shows that information about an object's velocity can be derived from the slope of its displacement–time curve or the area under its acceleration–time curve. If these processes were represented by Figure 1.6, y would be the displacement or acceleration and x would be time. The velocity could only be estimated with any accuracy if the data for y were collected for frequent and regular small intervals of x in those regions where y changed rapidly with x. Similarly, if the data were continuous, as produced by a pen recorder, the accuracy of the estimate of the velocity at points in time during the experiment would increase; this is because the estimate would be computed over smaller and smaller intervals of time. A fuller explanation is given in Chapter 7.

1.5 Engineering units

Statics and dynamics involve the study of real events, and like engineering in general, the data will be measured in real quantities. As such the data will require **units** of measurement such as metres for distance and kilograms for mass. Normally the units will conform to the SI system, using the following quantities:

Distance	metre	m
Mass	kilogram	kg
Force	newton	N
Velocity	metre per second	$m\,s^{-1}$
Pressure	newton per square metre	$N\,m^{-2}$

In many cases it is convenient to use larger or smaller units, then multiples of 1000 are used such as kilo (k = 10^3), mega (M = 10^6), milli (m = 10^{-3}). Intermediate powers of 10 are avoided.

It is often necessary in using equations to multiply or divide quantities involving units. This is acceptable and a new unit is usually produced. Hence metres (m) may be divided by seconds (s) to give $m\,s^{-1}$. It is not possible, however, to add or subtract quantities having different units. Hence metres plus seconds has no meaning. This is a useful check in all calculations; it ensures that units are compatible with the calculation being carried out.

➤ **1.6 Algebra**

In order to carry out calculations in engineering it is necessary to master the processes of algebra. Statements of physical laws, the equations for these laws and graphs representing them, all are usually cast in algebraic form. It is essential to understand how to manipulate algebraic expressions, to transpose and solve equations, and to derive information from graphs.

Statements of physical laws and other relationships can be quite lengthy and complex. Restatement of these laws in algebraic form is a form of shorthand, and gives great clarity to the relationship being described. Also, in the calculating process, algebra is a generalized form of arithmetic, and shows that the calculation applies to *all* numbers and not just the *particular* set of numbers being used at that time.

The processes of algebra are summarized below. For a full treatment of each item, reference should be made to one of the books listed in the further reading.

Symbols and notation

$$A + A = 2A$$

$$A \times A = A^2$$

$$A \times B = AB$$

Addition and subtraction

$$A + B = B + A$$

$$(A + B) + C = A + (B + C)$$

$$A - B = -B + A$$

Multiplication

$$A \times B = B \times A$$

$$A \times (B + C) = AB + AC$$

Division

$$(A + B)/C = A/C + B/C$$

Expansion

$$(A + B) \times (C + D) = A(C + D) + B(C + D)$$
$$= AC + AD + BC + BD$$
$$(A + B) \times (A + B) = A(A + B) + B(A + B)$$
$$= A^2 + 2AB + B^2$$
$$(A + B) \times (A - B) = A(A - B) + B(A - B)$$
$$= A^2 - B^2$$

Indices

$$A^n = A \times A \times A \ldots n \text{ times}$$

$$A^m = A \times A \times A \ldots m \text{ times}$$

$$A^n \times A^m = A^{(n+m)}$$

$$A^n / A^m = A^{(n-m)}$$

$$A^{-n} = 1/A^n$$

Equations

If $\quad\quad A = 2B$

then $\quad A + C = 2B + C$

and $\quad A - C = 2B - C$

and $\quad A \times D = 2B \times D$

and $\quad A/D = 2B/D$

All equations may be rearranged by operating on both sides of the equation identically. This is the balancing principle for equations. The intention of such operations is to produce a solution to the equation in which an unknown quantity is given the correct numerical value.

Formulae are often expressed as equations:

$$F = ma$$

$$\sigma = PL/AE$$

In these equations it may be useful to rearrange the algebra to give the **subject** of the formula as a different quantity. From the two examples above, rearrangement could give

$$a = P/M$$

$$E = PL/A\sigma$$

The method of rearrangement follows the same balancing principle for equations.

Simultaneous equations occur when two unknowns make up two equations. These may be solved by arranging one unknown as the subject of a formula, then substituting this in the second equation:

If $\quad A + B = 10$

and $\quad 2A - B = 8$

then from the first equation $B = (10 - A)$, which may be substituted in the second equation to give a result.

$$2A - (10 - A) = 8$$

$$2A + A = 8 + 10$$

$$3A = 18$$

$$A = 6$$

hence $\quad B = 10 - 6 = 4$

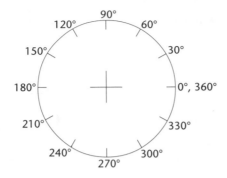

Figure 1.7   There are 360 degrees in a circle.

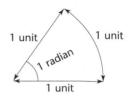

Figure 1.8   A unit arc subtends an angle of one radian.

Such solutions may be found for more sets of equations and unknowns, providing the number of equations and the number of unknowns are equal. For larger sets of equations and unknowns a knowledge of matrix algebra is an advantage, and practice with computer-based solutions is invaluable.

➤ 1.7   Trigonometry

Constructing any triangle will demonstrate that the sides and angles have some relationship with each other. The mathematics of the relations of angles and lengths of sides of triangles is known as trigonometry.

Lengths of sides of triangles, polygons, etc., are measured in units of **metres** (m) as the SI unit, but commonly mm or km will also be used. Angles are usually measured by **degrees**, where a full circle is 360° (Figure 1.7). An alternative, and often more useful method of angle measurement is the use of the **radian** in which one unit of arc is subtended by one unit of angle (the radian) when the arc radius is also one unit (Figure 1.8). Other methods of angle measurement exist but are of little importance in engineering.

It is important when using angles to make clear where the measurement of the angle is started, i.e. what is the baseline or zero position. For trigonometric functions the baseline is the *x*-axis, with angles measured anticlockwise from this line. This is known as the **polar** angle (Figure 1.9) and all trigonometric functions are given on this basis. Another alternative reference method is used in map reading, etc., where the angle is measured clockwise from the north direction and is known as the **bearing**.

There are three basic trigonometric functions which are related to the angles and side lengths of a right-angled triangle (Figure 1.10). Two sides are known as **adjacent**

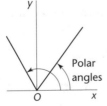

Figure 1.9   Polar angles are measured anticlockwise from a baseline.

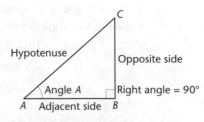

Figure 1.10   Nomenclature for a right-angled triangle.

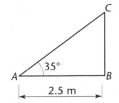

Figure 1.11 In this triangle the opposite side can be found using tan 35° and the hypotenuse using cos 35°.

to or **opposite** one of the acute angles, and the third side, opposite the right angle, is known as the **hypotenuse**.

Sine of angle A = sin A = opposite/hypotenuse

Cosine of angle A = cos A = adjacent/hypotenuse

Tangent of angle A = tan A = opposite/adjacent

For a further treatment of these functions, or ratios, reference should be made to one of the books listed in the further reading.

Each of the above trigonometric ratios has a reciprocal function: 1/sine is known as cosecant, 1/cosine as secant and 1/tangent as cotangent. But these ratios are not given on hand calculators and are generally best avoided; it is better to use 1/sine rather than cosecant.

Trigonometry is needed in much of statics and dynamics to obtain information from drawings or sketches, and to find solutions to diagrams. The most common requirement is the solution of the right-angled triangle (Figure 1.11), and in fact more complex shapes can be solved by reduction to a number of such triangles (Figure 1.12).

If the length of the base (Figure 1.11) is given as 2.5 m, and angle A is 35°, then

Height $BC = 2.5 \times \tan 35°$

$= 1.75$ m

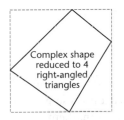

Figure 1.12 Complex shapes usually require simplification.

This is deduced from the ratio (given above) that tan A = opposite/adjacent.

In addition

Hypotenuse $AC = 2.5/\cos 35°$

$= 3.05$ m

This is deduced from the ratio that cos A = adjacent/hypotenuse.

In a similar way the more complex diagrams (Figure 1.12) can be solved by use of the right-angled triangle. Such a triangle can only be solved when two quantities are known, as well as the location of the right angle. These two quantities may be sides or angles, but not just two angles. If it is possible to draw the diagram from the information given, it is also possible to find all other quantities by use of the trigonometric ratios.

More complicated solutions of triangles not containing a right angle are available by sine rule and cosine rule. Figure 1.13 shows such a triangle in which the sine rule gives

$$\frac{a}{\sin A} = \frac{b}{\sin B} = \frac{c}{\sin C}$$

The cosine rule gives

$$a^2 = b^2 + c^2 - 2bc \cos A$$

$$b^2 = a^2 + c^2 - 2ac \cos B$$

$$c^2 = a^2 + b^2 - 2ab \cos C$$

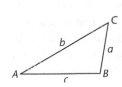

Figure 1.13 Scalene triangles are solved using sine and cosine rules.

These rules may be useful, but in most problems in statics and dynamics simpler

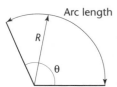

Figure 1.14 Arc length equals $R\theta$ where θ is measured in radians.

solutions are obtained by reduction to right-angled triangles (Figure 1.12). This concept is extended to the reduction of problems to components in the x and y directions, for example (Section 3.4).

In figures involving circles or arcs of circles it is often necessary to use the relationship between arc length, circle radius (R), and angle subtended (θ). From the definition of the radian unit of measurement at the beginning of this section, it is possible to derive the relation

Arc length $= R\theta$

provided θ is measured in radians (Figure 1.14). For a complete circle the arc length is equal to the circumference and is $2\pi R$. Hence the angle subtended (for a full circle) is 2π in *radians*. This gives rise to the relationship

$360° = 2\pi$ radians

Much of engineering involves very small angular movements, and many methods of calculation are summation techniques which add up these small increments. It is sometimes useful therefore to simplify formulae, etc., by including $\sin\theta$ or $\tan\theta$ as equal to θ, when θ (in radians) is very small. This useful simplification is often given in the form

$\sin\theta$ (or $\tan\theta$) tends to θ (if θ is small)

➤ 1.8 Calculus

Section 1.2 showed that many physical laws can be expressed as graphs. Section 1.3 explained how it is often necessary to perform calculations based on these graphs. Slopes and areas (Figures 1.4 and 1.5) might be required. This kind of mathematics is known as calculus, and may be found in one of the books listed in the further reading.

Differential calculus is a method of finding the slope of a graph, that is the rate of change of a given quantity. It measures the slope of a graph by the tangent to the curve at any point.

Integral calculus is a method of finding areas under graphs, that is the sum of quantities undergoing change. It measures areas by reducing them to infinitesimal units and summing the units.

If some measure, y, of the behaviour of an object can be represented by a curve, such as shown in Figure 1.5, and that curve can be described exactly by an equation in x, then y is said to be a function of x. As will be seen in Chapter 8, the differential of a function is associated with its slope, and the integral of a function is associated with the area under the curve it describes. Differentials and integrals are connected by a relationship of a reciprocal nature. The differential of a function describing the displacement of an object with time is associated with its velocity. Conversely, the integral of a function that describes the variation of its velocity with time is associated with its displacement.

If the behaviour of an object, such as its variation of displacement with time, can be represented in the form of an equation, then other variables relating to it, such as velocity and acceleration, can be calculated exactly at any time for which the equation is valid, using the methods of calculus. Similarly, if it can be represented by approximate curves drawn through experimental data, then other variables can be approximately calculated from the estimated slopes of the curves, and the areas under those curves. It also happens that many natural laws can be formulated in terms of differentials; this makes calculus a powerful tool in scientific and engineering calculations.

PART I Statics

Part I deals with statics, which is primarily concerned with the behaviour of forces acting on objects at rest. It is therefore of prime importance when considering the stability of buildings and similar large structures, but it is of equal significance when considering very small components. The subject of statics usually assumes a state of rest for the object or component, but it has equal meaning for states of uniform motion.

2 Vectors and definitions

2.1 Vectors

Figure 2.1 Vectors have magnitude and direction: 20 km in the wrong direction from Newcastle will not take you to Sunderland.

Some quantities may be fully described by giving their magnitude only, together with some meaningful units. For example we may give a time of 5 min, or a temperature of 10 °C. We may define a mass as 500 kg or a speed as 30 m s^{-1}. All these are **scalar** quantities.

By contrast, some quantities are not adequately described using magnitude alone, but require a direction in order to be fully defined. These are **vector** quantities. For example, we may give a movement (or linear displacement) as 20 km south-east (Figure 2.1), or a force as 10 N upwards. Velocity too, as distinct from speed, is usually given a direction, for example, 60 m s^{-1} horizontally.

A vector is a quantity having both **magnitude** and **direction**.

This chapter uses movements (linear displacements) to explore the properties of vectors and the rules governing their use. Then Chapter 3 will apply these rules to force vectors.

2.2 Vector addition

Figure 2.1 shows a movement vector of 20 km SE, illustrated (approximately) by a journey from Newcastle upon Tyne to Sunderland. A second movement vector of 20 km SW represents an extension of our journey from Sunderland to Durham (Figure 2.2). The overall effect of adding the two movements has the same result as a single movement of about 28 km S – Newcastle to Durham direct. Vector addition may therefore be simply achieved by drawing the vectors to scale as on a map (Figure 2.2). This is the graphical method of vector addition.

It may be seen (Figure 2.3) that the order in which the addition is carried out does not affect the end result or **resultant**. It will, in this case, affect what is seen on the journey, but not the destination. For the purposes of this book, only the resultant is

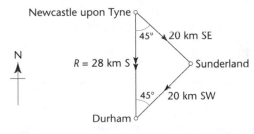

Figure 2.2 Two routes from Newcastle to Durham and their vectors.

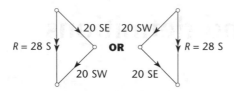

Figure 2.3 The end result does not depend on which way the vectors are added but the route does change. In this case you always get to Durham but you may not go via Sunderland.

required. In some more advanced aspects of engineering the order of addition can become important.

Using the same analogy of movements it is clear that any number of vectors can be added. Four movements (Figure 2.4(a)) or five vectors (Figure 2.4(b)) can be added using a graphical method as before to obtain a resultant.

Figure 2.4(b) introduces the principle of subtraction, as the movements of 14 km N and 14 km S cancel each other out. It is true to say that addition of 14 km S is the same as a subtraction of 14 km N.

Laying out all movements on a map (Figure 2.4(a)) may prove somewhat lengthy, and a shorter method of representing them at one point may be adopted (Figure 2.5). Figure 2.5(a) gives all the movements required by Figure 2.4(a), Figure 2.5(b) all those required by Figure 2.4(b). The order in which the movements are made has been eliminated. Alternatively, the vectors could be listed as follows:

Figure 2.4(a)	Figure 2.4(b)
14 km E	20 km SE
14 km S	14 km N
14 km S	14 km S
14 km W	20 km SE
28 km S (resultant)	28 km W
	28 km S (resultant)

If the addition of two vectors (Figure 2.2) can be achieved by drawing, calculations are also possible. Functions of the angles would be needed as in basic trigonometry (Section 1.7). In Figure 2.2 we have

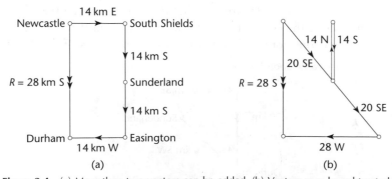

(a) (b)

Figure 2.4 (a) More than two vectors can be added. (b) Vectors may be subtracted: see how the two journeys of 14 km cancel each other out.

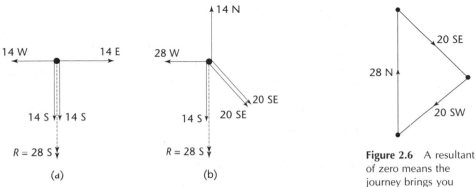

Figure 2.5   A shorter notation: (a) for Figure 2.4(a) and (b) for Figure 2.4(b).

Figure 2.6   A resultant of zero means the journey brings you back to your start.

Resultant $R = 20 \cos 45° + 20 \cos 45°$

$= 28.3 \text{ km}$

The addition of several movements may in some cases give a resultant of zero (Figure 2.6), i.e. the journeys bring the traveller back to his or her starting point. The end result is zero movement, and the graphical plot or diagram forms a closed figure. The closure of the diagram therefore indicates a sum to zero. Conversely, a zero resultant would require the diagram to be closed.

2.3   Vector components

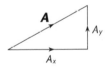

Figure 2.7   Vectors may be separated into components.

The addition of two vectors to give a single resultant suggests the reverse is true: any vector may be separated into two **components**. More than two components are possible but this is rarely useful. It is generally convenient to arrange the components at 90° (x and y directions) but this is not essential. Vector A in Figure 2.7 may be separated into two components A_x and A_y in the x and y directions. For movement vectors the components may be expressed in the directions N and E.

Example 2.1

Express each of the following movements D as components in the directions north and east:

(a) 18 km NW (north-west)

(b) 12 km SSE (south-south-east)

(c) 50 km 60° E of N (east of north)

Solution   Figure 2.8 sets out each movement together with the angles involved and the components. A graphical solution may be accomplished by drawing the movements to some scale and measuring the components. Alternatively, calculation may be used.

(a) $D_N = 18 \cos 45° = 12.7 \text{ km}$

$D_E = -18 \sin 45° = -12.7 \text{ km}$

The negative sign indicates the direction is west.

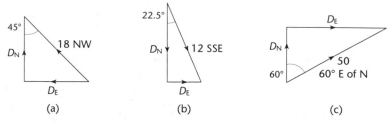

Figure 2.8 See Example 2.1.

(b) $D_N = -12 \cos 22.5° = -11.1$ km

$D_E = 12 \sin 22.5° = 4.6$ km

The negative sign indicates the direction is south.

(c) $D_N = 50 \cos 60° = 25.0$ km

$D_E = 50 \sin 60° = 43.3$ km

➤ **Example 2.2**

Express each of the following displacements **D** as components in the x and y directions.

(a) 2 km at 10° above the horizontal

(b) 30 mm at 20° anticlockwise from vertical downwards

(c) 2.5 m at 120° anticlockwise from the x-axis.

Solution As in Example 2.1, the solution may be obtained graphically or by calculation. Figure 2.9 shows each displacement together with the angles involved and the components.

(a) $D_x = 2 \cos 10° = 1.97$ km

$D_y = 2 \sin 10° = 0.35$ km

(b) $D_x = 30 \sin 20° = 10.3$ mm

$D_y = -30 \cos 20° = -28.2$ mm

The negative sign indicates a negative y component.

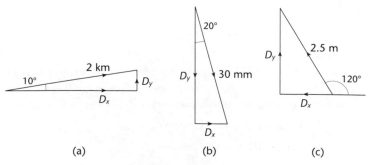

Figure 2.9 See Example 2.2.

(c) $D_X = 2.5 \cos 120° = -1.25$ m

$D_Y = 2.5 \sin 120° = 2.17$ m

Using the complete angle as given (anticlockwise from the X-axis) produces the correct signs for each component.

2.4 **Vector description**

The consideration of vector components and the two examples above show that a vector may be described fully by either of two methods.

Description 1   A vector has magnitude and direction and may therefore be described by a quantity (such as distance or length) and an angle from some reference line (such as north or the x-axis). When using a magnetic compass the angle is measured clockwise from north and it is called the **bearing** (Figure 2.10(a)). In coordinate geometry the angle is measured anticlockwise from the x-axis (polar coordinates) and is referred to as the **polar angle** α (Figure 2.10(b)). The advantage of using this method is that component signs are given automatically (Figure 2.9(c) and the example 2.2). For this method the x and y components are

$$D_x = D \cos \alpha$$

$$D_y = D \sin \alpha$$

Description 2   A vector may also be described by two components, generally at 90° (orthogonal). In map reading these two directions are east and north, and map references are given by

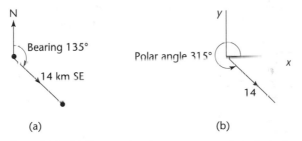

(a)   (b)

Figure 2.10   (a) Compass bearings are measured clockwise from north. (b) Polar angles are measured anticlockwise from the x-axis.

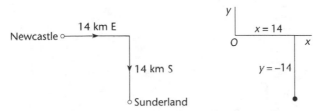

Figure 2.11   Grid references on a map use eastings and northings; Cartesian coordinates in the plane use x and y axes.

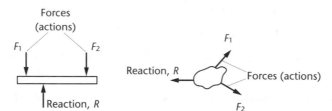

Figure 2.12 Actions are resisted by reactions.

eastings and northings; in coordinate geometry the directions are the x and y axes. (Figure 2.11).

These two methods give valid descriptions of a vector and in each case two quantities are essential.

In some cases a further piece of information may be desirable to give **location** to a vector. For example, Figures 2.1, 2.3 and 2.4 all used vectors, but located them by reference to Newcastle upon Tyne. The principles for adding the vectors did not depend on that location, and would be exactly the same somewhere else. The location of vectors relative to each other becomes important when considering force vectors which are not concurrent (see Section 3.7).

➤ 2.5 Forces

A force has both magnitude and direction and is therefore a vector. A force may result from particle attraction, such as gravity, a chemical action or explosion as in an internal combustion engine, or from human muscle power. A force is either a pull (tension) or a push (compression), and these two actions may be designated by a positive/negative sign convention. Forces are sometimes known as **actions** and forces which resist them and maintain stability as **reactions** (Figure 2.12). For example, the

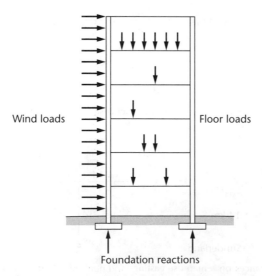

Figure 2.13 In a building the foundation reactions oppose the actions of wind loads and floor loads.

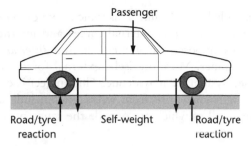

Figure 2.14 In a car the reactions between the tyres and the road oppose the car's weight, the weight of the passengers and other forces such as braking forces.

forces on a building (wind, floor loads, etc.) are actions which are resisted by reactions at the foundations (Figure 2.13). The forces on a vehicle (self-weight, passengers, braking, etc.) are resisted by reactions between the tyres and the road (Figure 2.14).

To affect an object a force must generally be in contact with that object, although some forces, such as gravity, may be inherent in the object. Structures and machinery are subjected to complex systems of forces, and for easy solution they are divided into parts or members (see Chapter 4). It is important to visualize each part and the forces to which it is subjected, which are applied at points of contact only. For example, a timber joist in a house has forces applied to it by the floorboards (at the points of contact) and is supported by reactions from the walls (at the points of contact).

A force must in general have **location** relative to other forces on a member. In the specific case of forces all acting at one point (concurrent forces) no further definition of location is required (see Section 3.5). In the general case, however, forces may act at different points on a member and their relative location must be defined. For example, forces on the building (wind, floor loads, foundation reactions) must be given a location relative to each other (Figure 2.13).

Although in practice forces must have a point of contact with the member considered, and hence a specific location, it is the **line of action** rather than the point of application that is important. A force is said to be **transmissible** when it has the same effect anywhere along its line of action (Figure 2.15). This property applies only as far as the addition and balance of forces is concerned, and must be used with caution when calculations of stress are required.

As for vectors, forces may be added and subtracted (Section 2.2) and broken into components (Section 2.3) by the rules already defined. For forces a sum to zero has a

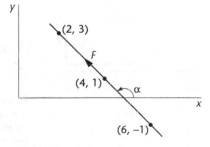

Figure 2.15 A transmissible force has the same effect anywhere along its line of action.

special meaning and is called **equilibrium**. This topic is explored in detail in Chapter 3.

Forces are sometimes described as **external** forces and **internal** forces. In general *external* refers to forces imposed on a member and *internal* to resulting forces generated within the member. More precisely, external forces are those acting from outside upon any closed system we may define. This subject is considered in detail in Chapter 4. Internal forces may arise from stress and include concepts of bending moment and shear force, all of which are outside the scope of this book.

➤ 2.6 Moments and couples

Figure 2.16 A moment is the turning effect of a force about a point.

$$M_o = Fp$$

The moment of a force is the product of the magnitude of the force and the perpendicular distance of its line of action from the point about which moments are being calculated (Figure 2.16). It is the turning effect of the force about a point.

$$M_o = Fp$$

The simple lever system shown in Figure 2.17(a) and the seesaw of Figure 2.17(b) illustrate the principle of the balance of moments as well as forces. A zero sum or equilibrium is as important for moments as it is for forces (see Section 3.7).

A **couple** is the special case of two equal parallel forces (in the same plane) acting in opposite directions. Figure 2.18 depicts a couple and it may be shown that the moment generated is independent of the distance p, i.e. the moment has a constant value Fd about any point in its plane.

$$M_o = F(p + d) - F(p) = Fd$$

which is independent of the position of O.

Recap

- A vector has magnitude and direction.

- Vectors may be added together or separated into components by drawing or by trigonometry.

- Forces are vectors having magnitude and direction.

- Moments combine force and perpendicular distance.

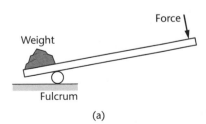

(a)

$F_1 p_1$ and $F_2 p_2$ balance

(b)

Figure 2.17 Examples of moments: (a) simple lever and (b) seesaw.

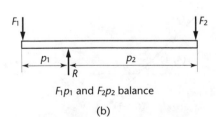

$$M_o = Fd$$

Figure 2.18 A couple is a special case of two equal parallel forces; the moment depends only on distance d, not distance p.

2.7 Problems

Give your answers by drawing to two significant figures and by calculation to three significant figures, with correct units.

1. Find the vector sum (resultant) of the movements

    5 km E
    6 km S
    20 km SW

   by (a) drawing and (b) calculation.

2. Find the vector sum of the movements

    5 km W
    6 km N
    20 km NE

   by drawing or by calculation. Define a relationship between the vectors given in Problem 1 and those in Problem 2.

3. Find the vector sum of the movements

    20 km SW
    5 km E
    6 km S

by drawing or by calculation. Note that the answers in Problems 1 and 3 are the same.

4. Find the movement vector **P** such that the following four movements sum to zero:

    5 km E
    6 km S
    20 km SW
    P

   What is the relationship of the vector **P** with the resultant found in Problem 1?

5. Describe each of the following vectors:

   Using (a) components in east and north directions
       (b) distance and polar angle

    5 km E
    6 km S
    20 km SW

6. By summing the components (Problem 5(a)) in each direction, derive the sum of the vectors.

7. Using distance and polar angle (Problem 5(b)) derive a method of summing the vectors.

3 Equilibrium

Section 2.5 showed that force is a vector quantity having both magnitude and direction. Vector addition must be employed when combining forces and a summation to zero has special significance. Forces may be actions or reactions resulting from several effects such as gravity. The moment of a force (Section 2.6) is the product of its magnitude and its perpendicular distance.

In this chapter we shall define the interaction of forces more carefully and examine in particular the conditions for equilibrium.

3.1 Newton's laws

Three laws of motion are generally known as Newton's laws of motion. These laws have equal significance for bodies at rest and form a basis for the study of statics. They are statements of observed events, and are therefore fundamental physical laws.

Law 1 *A particle remains at rest, or continues at uniform velocity in a straight line, unless acted on by a force.*

The force considered here is a general term which could mean the net effect of a system of forces, i.e. the vector sum of a number of forces. This resultant could be a force or a moment or both (see Chapter 2). This law may be used to derive the principle or 'condition' of equilibrium: for a body to be in equilibrium the forces acting on it must sum to zero.

Law 2 *The acceleration of a particle is proportional to the force acting on it and is in the direction of the force.*

As before the force is general and includes any moments. This law does not apply in cases of equilibrium, but is used when considering certain effects of motion in Parts II and III.

Law 3 *The forces of action and reaction between particles are equal in magnitude, opposite in direction, and are collinear.*

This law has particular application to systems of members which are common in engineering. The transfer of forces between members requires that action and reaction are equal (magnitude), opposite (direction) and collinear. This condition is demonstrated later in this chapter and in Chapter 4.

3.2 Force summation

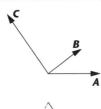

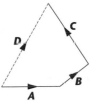

Figure 3.1 Three vectors may be summated to give a resultant: $A + B + C = D$.

We have already seen that vector quantities may be added together, and how this operation may be represented graphically by a drawing.

Movement vectors and force vectors obey the same rules of addition. Section 2.2 showed that the order in which vector addition is carried out is unimportant as far as the end result is concerned. This is equally true for the addition of force vectors. But note that for more advanced study the order becomes important in particular situations.

Vector addition may be carried out graphically or by calculation, as described in Section 2.2, and it may be used for any number of forces.

Three forces (vectors) **A**, **B**, and **C** acting on a body may be summated (Figure 3.1) and give a resultant force **D**. In the special case of two forces **A** and **B** only being summated as **E** (Figure 3.2), a parallelogram may be used as a graphical representation. This is known as the parallelogram of forces.

In practice, drawing the forces to some convenient scale would produce a satisfactory answer, or a knowledge of trigonometry (Section 1.7) would allow an answer by calculation.

3.3 Force components

The summation of forces using a parallelogram suggests that the inverse is true: any force (vector) may be separated into components, i.e. force **E** may be separated into components **X** and **Y** (Figure 3.3). It is convenient to arrange the components at 90° (x and y directions) but this is not essential.

A force may be separated into any number of components, but this is generally of no value. However, the use of orthogonal components is convenient as demonstrated in Section 2.3.

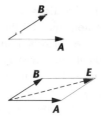

Figure 3.2 Two vectors may be summated graphically using a parallelogram.

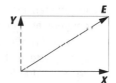

Figure 3.3 The opposite of summation is to separate a vector into components, in this case along the X and Y directions.

➤ **Example 3.1**

Find the components in the X and Y directions of each of the forces **A**, **B** and **C**, shown in Figure 3.4.

Solution

A = 12 kN

45°

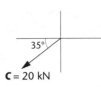

B = 10 kN

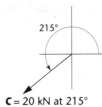

35°

C = 20 kN

Figure 3.4 See Example 3.1.

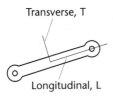

215°

C = 20 kN at 215°

Figure 3.5 See Example 3.1.

Force **A**   component $A_x = 12 \cos 45°$
$$= 8.49 \text{ kN}$$
component $A_y = 12 \cos 45°$
$$= 8.49 \text{ kN}$$

Force **B**   component $B_x = 10 \cos 0° = 10 \text{ kN}$
component $B_y = 10 \cos 90° = 0$

This result confirms what may be self-evident, that a force has zero component at 90° to its direction.

Force **C**   component $C_x = -20 \cos 35°$
$$= -16.38 \text{ kN}$$
component $C_y = -20 \sin 35°$
$$= -11.48 \text{ kN}$$

These components are counted as negative because their directions are in the negative x and negative y directions. An alternative way of defining the vector **C** may be used (as shown in Section 2.3) which will give the negative sign directly. Here the vector angle is defined from the positive x-direction, measuring anticlockwise (Figure 3.5). The components of **C** may now be expressed as follows:

Force **C**   component $C_x = 20 \cos 215°$
$$= -16.38 \text{ kN}$$
component $C_y = 20 \sin 215°$
$$= -11.48 \text{ kN}$$

In general, components of force in directions at right angles to each other are the most convenient. These orthogonal directions may be the x and y axes of an arbitrary reference system. An alternative system may use horizontal and vertical directions, which are clearly appropriate to buildings. Longitudinal and transverse reference directions (Figure 3.6) are equally suitable and have particular use for members of machines.

The commonest error when expressing a force (vector) as its components is to 'resolve' a force in a particular direction. This gives one component only (in the given direction) and commonly results in the second component being forgotten. It is best to avoid all ideas of 'resolving' forces, and to grasp the concept of expressing a force as two components. This retains a proper appreciation of a force as a vector. In answers to questions it is equally satisfactory to give a force by magnitude and direction, or by two components in specified directions.

Transverse, T

Longitudinal, L

Figure 3.6 Vectors may be separated into longitudinal and transverse components.

➤ **3.4 Force summation by components**

The separation of a force into its components suggests an alternative method of force summation. We have seen in Section 3.2 that force vectors may be summated graphically. It is possible, as shown in Chapter 2, to carry out this operation by drawing or by using geometry and the functions of the appropriate angles. An alternative and more systematic method is to summate by use of the force components. In Example 3.1 forces A, B, C were separated into their components. The sum of these forces is clearly the sum of the components, so

if $A + B + C = D$, the resultant force vector

and $A = A_x + A_y$, etc.

then $D_x = A_x + B_x + C_x$

and $D_y = A_y + B_y + C_y$

$$D = D_x + D_y \text{ (vector addition)}$$

▶ Example 3.2

Find the sum of the forces shown in Figure 3.4 using the components of force already calculated.

Solution   If D is the vector summation (Figure 3.7) then

$$D_x = 8.49 + 10.00 - 16.38$$

$$= 2.11 \text{ kN}$$

$$D_y = 8.49 + 0 - 11.48$$

$$= -2.99 \text{ kN}$$

$$D = \sqrt{(2.11^2 + 2.99^2)}$$

$$= 3.66 \text{ kN}$$

$$\alpha = \tan^{-1}(2.99/2.11)$$

$$= 54.8°$$

or $D = \mathbf{3.66}$ **kN at 305.2°** (Figure 3.8)

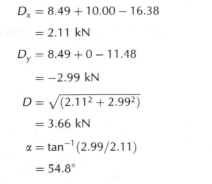

Figure 3.7 See Example 3.2.

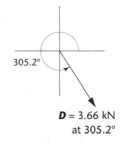

Figure 3.8 See Example 3.2.

$D = 3.66$ kN at 305.2°

▶ 3.5 Equilibrium of concurrent forces

Forces which act at a point are described as concurrent. In Section 3.7 we shall consider the effects of forces being non-concurrent. In these sections we consider force vectors being in one plane – two-dimensional systems or **coplanar** forces. Section 2.2 showed that when a number of vectors were summated and the result was zero, a special case of no resultant movement or no resultant force was achieved. The case of no resultant force was described as the basic concept of equilibrium if the vector sum of the forces, $\sum F$ (the resultant), is zero. This may be expressed either analytically or graphically (Section 3.6).

Using the concept of components, as given in Section 3.4, the condition for equilibrium may be expressed as follows:

Sum of x-components = zero

Sum of y-components = zero

or $\sum F_x = 0$ and $\sum F_y = 0$

➤ **Example 3.3**

Show that the three concurrent forces given in Figure 3.9 are in equilibrium.

Solution

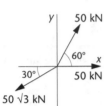

Figure 3.9 See Example 3.3.

$$\sum F_x = 50 + 50 \cos 60° + 50\sqrt{3} \cos 210°$$

$$= 50 + 25 - 75 = 0$$

$$\sum F_y = 0 + 50 \sin 60° + 50\sqrt{3} \sin 210°$$

$$= 0 + 43.3 - 43.3 = 0$$

Hence the condition for equilibrium is satisfied.

➤ **Example 3.4**

Find the magnitude and direction of the force **P** if the four concurrent forces shown in Figure 3.10 are in equilibrium.

Solution

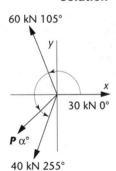

Figure 3.10 See Example 3.4.

$\sum F_X = 0$ for equilibrium

$30 + 60 \cos 105° + P \cos \alpha + 40 \cos 255° = 0$

$30 - 15.53 + P \cos \alpha - 10.35 = 0$

$P \cos \alpha = -4.12$ (3.1)

$\sum F_Y = 0$ for equilibrium

$0 + 60 \sin 105° + P \sin \alpha + 40 \sin 255° = 0$

$0 + 57.96 + P \sin \alpha - 38.64 = 0$

$P \sin \alpha = -19.32$ (3.2)

Dividing Equation 3.2 by Equation 3.1 gives

$\tan \alpha = P \sin \alpha / P \cos \alpha$

$= -19.32/-4.12 = 4.689$

$\alpha = 78.0°$ or $258.0°$

If $\alpha = 78.0°$ then Equation 3.1 gives

$P = -4.12/\cos 78°$

$= -19.8$ kN

If $\alpha = 258.0°$ then

$$P = -4.12/\cos 258°$$

$$= 19.8 \text{ kN}$$

Hence **P = 19.8 kN at 258°**

3.6   Polygon of forces

When the vector sum of a number of forces acting at a point is zero, the condition for equilibrium is satisfied.

Section 3.5 showed how this could be expressed using force components. The vectors may alternatively be expressed graphically by representing them as lines of appropriate length and direction. Summation of the vectors is represented by drawing a polygon as shown in Section 2.2. If the vectors sum to zero, the polygon will close and the resultant will be zero.

Forces **A**, **B**, **C**, **D** may be represented as in Figure 3.11 by lines of appropriate length. The summation is shown in Figure 3.12 as a polygon of forces (a quadrilateral in this case). Closure of the polygon indicates that the equilibrium condition is satisfied. Failure of the polygon to close would indicate a resultant force as shown in Section 2.2.

Different polygons are produced when the forces are summated in different orders but the resultant is always zero (Figure 3.13).

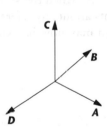

Figure 3.11   Four forces acting at a point.

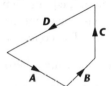

Figure 3.12   The summation as a polygon of forces – in this case, a quadrilateral.

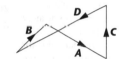

Figure 3.13   A different polygon is produced when the forces are summated in a different order, but it is still closed.

3.7   Equilibrium of non-concurrent forces

So far it has been assumed that all the forces act on a particle, i.e. at one point, and are therefore concurrent. In the more general case a system of forces (Figure 3.14) may not act through one point, and will therefore have a turning or moment effect relative to each other (see Section 2.6 for definitions). For such a non-concurrent system a resultant force and a resultant moment may be found. Any arbitrary reference point may be used to find the resultant, e.g. the point O. As before (Section 3.4) the resultant R may be found by the use of x and y components.

$$R = \sqrt{\left[\left(\sum F_x\right)^2 + \left(\sum F_y\right)^2\right]}$$

and $\tan \theta_R = \sum F_y / \sum F_x$

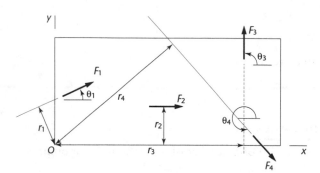

Figure 3.14 Non-concurrent forces may lead to a resultant moment as well as a resultant force.

And the resulting moment M_o about O is

$$M_o = \sum Fr$$

(a)

(b)

Figure 3.15 Two ways to represent the same resultant force and resultant moment at O: (a) a concurrent force plus a moment and (b) a non-concurrent force arranged at the appropriate distance to generate M_o.

where r is the radius or lever arm from O perpendicular to the line of action of each force. Note that the appropriate sign convention takes clockwise moments positive, and anticlockwise moments negative.

The resultants of the system (Figure 3.15(a)) are R and M_o, and the resultant force R passes through O and has zero moment about O. Alternatively the resultant force R may be arranged to act at such a distance from O as to give the moment M_o. In this case the force R acting at a distance r_0 from O defines the resultant (Figure 3.15(b)).

As in Section 3.5, equilibrium is the special case when the system of forces sums to zero, but for non-concurrent forces the moments about any reference in the plane must also sum to zero. The condition of equilibrium may now be expressed as follows:

Sum of x-components = zero

Sum of y-components = zero

Sum of moments about any point = zero

or $\quad \sum F_x = 0, \quad \sum F_y = 0, \quad \sum M_o = 0$

There are only *three* conditions of equilibrium for non-concurrent coplanar forces. These three conditions give rise to three equations, and no more. Summation of forces about a different set of axes, or summation of moments about a different point may appear to produce additional equations. When examined, however, these prove to be merely dependent equations which could be derived from the original three.

Nevertheless, the three conditions of equilibrium may be combined in several different sets (Table 3.1) as appropriate to the problem. A, B and C are three points anywhere in the plane (for Set 4 they must not be in one straight line).

Table 3.1 Conditions of equilibrium

Set 1	Set 2	Set 3	Set 4
$\sum F_x = 0$	$\sum F_x = 0$	$\sum F_y = 0$	$\sum M_A = 0$
$\sum F_y = 0$	$\sum M_A = 0$	$\sum M_A = 0$	$\sum M_B = 0$
$\sum M_A = 0$	$\sum M_B = 0$	$\sum M_B = 0$	$\sum M_C = 0$

Example 3.5

Find the magnitude, direction and position of force P if the four forces are in equilibrium (Figure 3.16(a)).

Solution   Components in X-direction

$$P_X + 30 \cos 45° + 40 \cos 120° + 40 \cos 135° = 0$$

$$P_X = 27.1 \text{ kN}$$

Components in y-direction

$$P_y + 30 \sin 45° + 40 \sin 120° + 40 \sin 135° = 0$$

$$P_y = -84.1 \text{ kN}$$

$$P = \sqrt{(27.1^2 + 84.1^2)} = \mathbf{88.3 \text{ kN}}$$

$$\alpha = \tan^{-1}(-84.1/27.1) = \mathbf{287.9°}$$

$$r_1 = 0.6 \cos 30° = \mathbf{0.520 \text{ m}}$$

$$r_2 = 0.6/\cos 45° = \mathbf{0.849 \text{ m}}$$

Moments about point A

$$-P \times r_A + 40 \times 0.520 + 40 \times 0.849 = 0$$

$$r_A = \mathbf{0.620 \text{ m}}$$

The force P may be shown as in Figure 3.16(b).

It would be neater, and less liable to error, to cast these kinds of calculation in tabular form (Table 3.2).

The three conditions of equilibrium may be applied in the same way and the same answers should be obtained. When completing the column for moments in the table, it is necessary to adopt a sign convention: clockwise positive, anticlockwise negative.

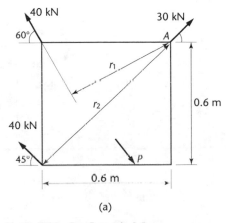

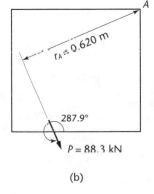

(a)   (b)

Figure 3.16   See Example 3.5.

Table 3.2 Tabular layout

Force, F	Component, F_x	Component, F_y	Radius, r	Moment, M_A
40	40 cos 135°	40 sin 135°	0.849	34.0
40	40 cos 120°	40 sin 120°	0.520	20.8
30	30 cos 45°	30 sin 45°	0	0
P	$P \cos \alpha$	$P \sin \alpha$	r_A	$-Pr_A$
	$\sum F_x = 0$	$\sum F_y = 0$		$\sum M_A = 0$

➤ **Example 3.6**

Find the magnitude, direction and position of force P if an anticlockwise moment of 10 kN m is added at point A to the plate examined in Example 3.5 (see Figure 3.17).

Solution   A moment of this nature is a couple (see Section 2.6) which has no net force in any direction, and has the same value at any point in the plane. So

$$P_x = 27.1 \text{ kN}$$

$$P_y = -84.1 \text{ kN}$$

$$P = 88.3 \text{ kN} \qquad \text{(all as previous solution)}$$

Moments about point A

$$- P \times r_A + 40 \times 0.520 + 40 \times 0.849 - 10 = 0$$

$$r_A = \textbf{0.507 m}$$

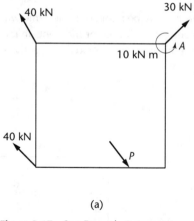

(a)

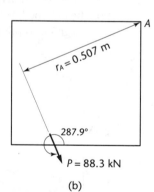

(b)

Figure 3.17   See Example 3.6.

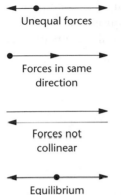

Unequal forces

Forces in same direction

Forces not collinear

Equilibrium

Figure 3.18   Four possible configurations for two forces.

If two forces acting on a rigid body are to be in equilibrium, all the following conditions must be met:

➤ The forces must be equal in magnitude.
➤ The forces must be opposite in direction.
➤ The forces must be collinear (which includes concurrent).

Figure 3.18 illustrates what happens when any one of these conditions is broken. The consequence is always a resultant force or moment, i.e. no equilibrium. This result, although perhaps self-evident, is of particular importance when considering pin-jointed frames (Chapter 6). All members of a pin-jointed frame are acted on by two forces, one at each end. Hence, for straight members, the forces have their direction along the line of the member.

3.9   Special case of three forces

If three forces acting on a rigid body are to be in equilibrium, all the following conditions must be met:

Sum of x-components $= 0$

Sum of y-components $= 0$

Must meet at a common point (be concurrent) or fulfil the conditions of Section 3.10

The third condition is illustrated in Figure 3.19, in which two of the forces are summated, reducing the system to two forces only. The conditions of Section 3.8 now apply, one of which is concurrency. An alternative to this condition is that the forces are parallel (Section 3.10).

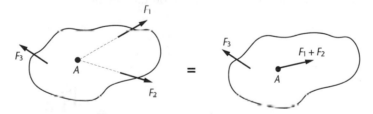

F_3 and $(F_1 + F_2)$ must satisfy the conditions in Section 3.9

Figure 3.19   For equilibrium of three forces acting on a rigid body, the X-components must sum to zero, the Y-components must sum to zero, and all three must be concurrent – they must meet at a single point – unless they are parallel.

➤ 3.10 Special case of parallel forces

If the three forces acting on a rigid body are in equilibrium, the condition of concurrency applies (Section 3.9). An alternative to this condition is that the forces may be parallel, and such a system is illustrated by the simple arrangements of lever and fulcrum, or beam and supports (Figure 3.20).

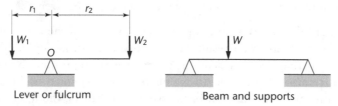

Figure 3.20 Equilibrium of parallel forces in a lever and fulcrum, and a beam and supports.

➤ 3.11 Centre of force

The fulcrum must be placed at the point of balance for the forces on the lever to be in equilibrium. In Figure 3.20 the fulcrum must be positioned at O so that

$$\sum M_o = 0$$

i.e. $\quad W_1 \times r_1 + W_2 \times r_2 = 0$

The position O may be described as the centre of force, with regard to forces W_1 and W_2. It is that point about which the moments of the forces sum to zero. Strictly speaking it is not so much a point as a line of action for a force. In some situations it is convenient to consider a force system as concentrated at this centre of force. Hence in Figure 3.20 the two-force system has a centre of force at O, and the four-force system in Figure 3.21 has a centre of force on the line of R.

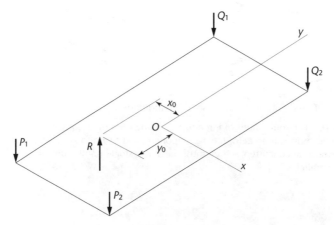

Figure 3.21 This four-force system has a centre of force on the line of R.

In the general case the centre of force is positioned at (x_0, y_0) with respect to some reference axes (Figure 3.21) such that

$$x_0 = \sum Fx / \sum F$$
$$y_0 = \sum Fy / \sum F$$

The concepts centroid of mass, centroid of area, etc., are similar, and the position (x_0, y_0) of a centroid relative to reference axes is defined by

$$x_0 = \sum Ax / \sum A$$
$$y_0 = \sum Ay / \sum A$$

Example 3.7

Four forces act on a roof structure as shown in Figure 3.22. Find the resultant force and the position of the centre of force.

Solution   Summate the forces in a vertical direction

$$R = 0.5 + 1.0 + 1.0 + 1.0 = \textbf{3.5 kN}$$

Take moments about O

$$Rx_0 = 0.5 \times 0 + 1.0 \times 2 + 1.0 \times 4 + 1.0 \times 6 = 12.0 \text{ kN m}$$
$$x_0 = 12.0/3.5 = \textbf{3.43 m}$$

Example 3.8

Find the position of the centroid of the area shown in Figure 3.23, giving its position relative to the x and y axes.

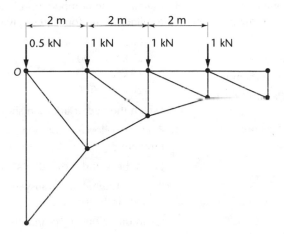

Figure 3.22 See Example 3.7.

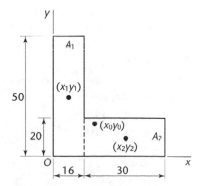

Figure 3.23 See Example 3.8.

Solution

$$\sum A = A_1 + A_2$$

$$= 16 \times 50 + 30 \times 20$$

$$= 800 + 600 = 1400 \text{ mm}^2$$

$$\sum Ax = A_1 x_1 + A_2 x_2$$

$$= 800 \times 8 + 600 \times 31 = 25\,000 \text{ mm}^3$$

$$\sum Ay = A_1 y_1 + A_2 y_2$$

$$= 800 \times 25 + 600 \times 10 = 26\,000 \text{ mm}^3$$

$$x_0 = 25\,000/1400 = \mathbf{17.9 \text{ mm}}$$

$$y_0 = 26\,000/1400 = \mathbf{18.6 \text{ mm}}$$

Recap

■ **Forces are vectors and may be summed using vector rules.**

■ **Forces may be separated into components using vector rules.**

■ **Forces are in equilibrium when their vectors sum to zero.**

■ **For non-concurrent forces to be in equilibrium the sum of their moments must also be zero.**

■ **Special conditions may be deduced when only two or three forces act.**

➤ 3.12 Problems

Give your answers by drawing to two significant figures and by calculation to three significant figures, with correct units.

1. Find the resultant of two tensile forces $2P$ and $3P$ acting on a particle if the angle between them is $60°$; obtain the solution (a) by drawing and (b) by calculation.

2. Find the components in the x and y directions of the following forces:

30 kN	135°
40 kN	310°
6 N	5°
−50 kN	215°
−120 N	270°

3. Calculate the components of each force to find the resultant of the four forces given below.

60 kN	30°
−20 kN	90°
−10 kN	240°
−20 kN	300°

   Sketch the vector addition showing the resultant.

4. Find the following components for the resultant in Problem 3:

   (a) in the x and y directions

   (b) longitudinal and transverse to a member whose axis is at 30°

   (c) in directions at 0° and 30°

   Sketch the resultant with each pair of components.

5. Forces of 5 N, 6 N, 7 N and 8 N act as shown in Figure 3.24 where the coordinates are in metres.

(a) Find the magnitude and direction of their resultant R.

(b) Find the line of application of their resultant R.

(c) Find the couple M that must be applied in the plane to make R pass through the origin.

(d) Find the required location of M.

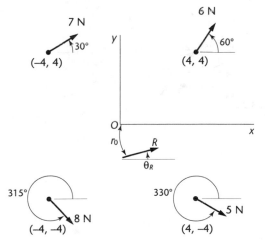

Figure 3.24

6. The three tensile forces below act at one point and are in equilibrium. Find the values of P and α.

10 kN, 30°
10 kN, 90°
PkN, α°

7. Four tensile forces acting at a point are in equilibrium. Find the values of P and Q.

100 N   30°
100 N   90°
PN   180°
QN   270°

8. Three forces P, Q and R act at a point and are in equilibrium. Find the ratios Q/P and R/P if the angles of the forces are as follows:

(a) 0°, 120°, 240° respectively

(b) 0°, 90°, 210° respectively

(c) 0°, 60°, 180° respectively

9. Find the forces P and Q (and R in Figure 3.25(e)) required for equilibrium in each of the lever systems shown in Figure 3.25.

(a)
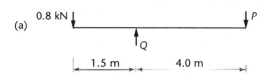

(b)
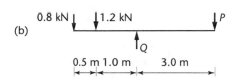

(c)
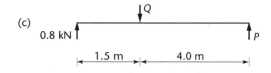

(d)
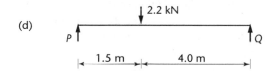

(e)
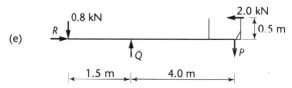

Figure 3.25

10. A triangular plate is acted on by the forces shown in Figure 3.26, and equilibrium is maintained by a single force P. Find the magnitude, direction and position of P.

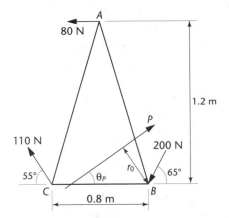

Figure 3.26

11. A plate is acted on by forces $2P$, $3P$, and $4P$ as shown in Figure 3.27. Equilibrium of the plate is maintained by the addition of force F at point A, and by the positioning of the $2P$ force. Find the magnitude and direction of force F, and the position of the $2P$ force.

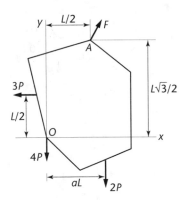

Figure 3.27

12. A circular plate shown in Figure 3.28 is acted on by forces of 10 N, 20 N, 16 N, 30 N and P. If the plate is in equilibrium under the action of these forces, find the magnitude and direction of P, and the intercept of its line of action on the y-axis.

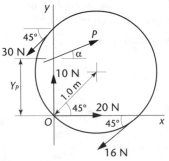

Figure 3.28

4 Force systems

The concepts of forces as vectors and their arrangement to give the condition known as equilibrium were set out in Chapters 2 and 3. These form the basis of all statics and hence of all structural analysis. In this chapter we define force systems more precisely, discover how they interact, and apply the conditions of equilibrium to them. It will then be possible to cope with the more complex structures examined in Chapters 5 and 6. And so that actual structures may be investigated, and the whole subject given reality, the nature of real forces will be examined.

4.1 Inclined planes

Any object supported on a horizontal plane experiences a vertical reaction from the plane (Figure 4.1). This reaction is equal to the weight of the object, it satisfies Newton's third law (Section 3.1) and it produces equilibrium (Section 3.8). But if the plane is inclined as shown in Figure 4.2, the reaction R is perpendicular to the surface of the plane and is often called the **normal reaction**. The forces W and R no longer produce equilibrium by themselves and a further force is required. In Figure 4.2 this is shown as a **friction force** F acting along the surface of the plane. If the angle of the plane is steep enough, this friction force may reach a maximum value of μR where μ is known as the coefficient of friction. For a fuller treatment of this topic see Chapter 11.

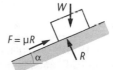

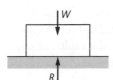

Figure 4.1 An object supported on a horizontal plane experiences a vertical reaction.

Figure 4.2 An object supported on an inclined plane experiences a reaction perpendicular to the plane, the normal reaction.

Figure 4.3 Sometimes a force is needed to maintain equilibrium. This may depend on the friction between the object and the plane, the angle of the plane and the shape of the object. The object's shape will affect the area of contact.

Figure 4.4 Normal reactions are perpendicular to surfaces in contact. The surfaces could be two spheres.

In cases where the angle of the plane is even higher, and friction cannot produce equilibrium, some other force is needed. Similarly, in cases of cylinders or spheres on the inclined plane (Figure 4.3) which experience negligible friction, some other force is also needed. A cable might provide such a force P as shown. Note that in both Figures 4.2 and 4.3 the special case of three forces will apply (Section 3.9) and the forces must meet at one point.

The normal reaction is perpendicular to the surfaces in contact (Figure 4.4). This leads to the concept of stable and unstable equilibrium shown in Figure 4.5. The cylinder or roller will return to the central position in Figure 4.5(a) after being given a small displacement. But in Figure 4.5(b) the roller remains in equilibrium in the new position following a displacement. In contrast, Figure 4.5(c) shows what appears to be equilibrium, but a small displacement leads to loss of equilibrium. This case is known as unstable equilibrium. All three cases are useful in engineering, but it must be clear which is likely to occur in any given situation. Stable equilibrium is desirable in structures. Unstable equilibrium, which must be avoided in structures, may be desirable in some control mechanisms.

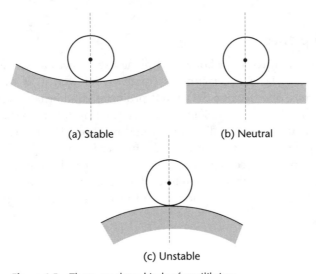

(a) Stable   (b) Neutral

(c) Unstable

Figure 4.5 There are three kinds of equilibrium.

➤ **Example 4.1**

A cube of weight W rests on a plane inclined at α to the horizontal. If the coefficient of friction is μ, and an additional force P must be applied via the cable to maintain equilibrium, find the value of P (Figure 4.6(a)).

Solution   If μR is the maximum friction force that occurs, the extra force P is needed to maintain equilibrium. For equilibrium, the four force vectors W, R, P and μR must sum to zero, i.e. the vector diagram (Figure 4.6(b)) must close. The diagram shows that the vectors are related by trigonometric functions.

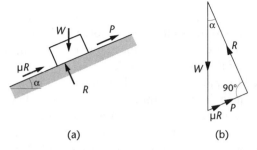

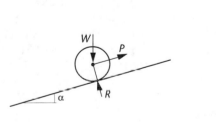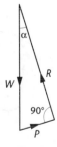

(a)   (b)

Figure 4.6 See Example 4.1.

$$\sin \alpha = (\mu R + P)/W \qquad (4.1)$$

$$\cos \alpha = R/W \qquad (4.2)$$

Rearranging Equation 4.1 gives

$$W \sin \alpha = (\mu R + P)$$

$$P = W \sin \alpha - \mu R$$

Rearranging Equation 4.2 gives

$$R = W \cos \alpha$$

> ## Example 4.2

If the cube in the previous example is replaced by a sphere, what is the new cable force?

Solution   The new arrangement of forces is shown in Figure 4.7 together with the vector diagram. As before, the diagram must close and the trigonometric function $\sin \alpha$ gives

$$\sin \alpha = P/W$$

hence

$$P = W \sin \alpha$$

Figure 4.7 See Example 4.2.

➤ 4.2 Pulley systems

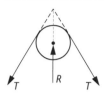

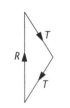

Figure 4.9 Equilibrium of the pulley requires a reaction at its axle.

Ropes and cables are used to transmit force as shown in Figure 4.8. The force is always tension; ropes and cables are considered incapable of transmitting compression. The force has direction along the line of the cable.

Pulleys may be used to change the force directions (and to allow the cable to move) without changing its magnitude. In these exercises the pulleys are assumed to be frictionless. Equilibrium of the pulley (Figure 4.9) requires a reaction – a force vector – at the axle of the pulley, so the vector diagram may be drawn as shown. The diagram commonly takes the form of an isosceles triangle (two sides of equal length).

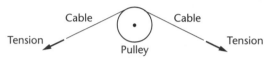

Figure 4.8 The force in a rope or cable is always tension.

➤ Example 4.3

Cable and pulley arrangements are used to lift a weight of 100 kN as shown in Figure 4.10. What is the cable tension T, and the reaction (vector) R at the axle of the pulley?

Solution

The pulleys shown in Figure 4.10(a) may be isolated as indicated; equilibrium of each pulley may be stated.

Equilibrium of the lower pulley

$$\sum F_y = 0$$

$$T + T - 100 = 0$$

hence

$$T = 50 \text{ kN}$$

Equilibrium of the upper pulley

$$\sum F_y = 0$$

$$R - 50 - 50 = 0$$

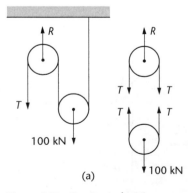

(a)

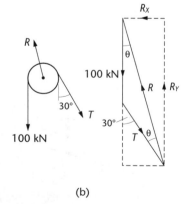

(b)

Figure 4.10 See Example 4.3.

hence

R = 100 kN vertically

The pulley shown in Figure 4.10(b) is in equilibrium, so the vector diagram must close. The cable is continuous, so the tensions are equal and T is 100 kN. Trigonometric functions within the vector diagram give the following values.

$\sin 30° = R_x/T$

$\cos 30° = (R_y - 100)/T$

hence

$R_x = 50$ kN

and

$R_y = 187$ kN

$R = \sqrt{(50^2 + 187^2)}$

= 193 kN at 15° to the vertical

The vector diagram is an isosceles triangle in which θ is 15°.

4.3   Force systems

In this chapter, and in Chapter 3, an arrangement or system of forces has been accepted as those acting 'at a point' or 'on an object'. In Example 4.3 the forces assumed to be in equilibrium were the forces on the lower pulley, along with the forces on the upper pulley, as shown in Figure 4.10(a). By isolating the forces acting on a specific part of the pulley arrangement, we have defined a **system** of forces.

A system of forces may in fact be defined by an arbitrary boundary arranged for our own convenience, and determined to some extent by our experience. The boundary of the system must close, and any forces crossing this boundary are deemed to enter or leave the system. These forces, but only these forces, must be included in any vector additions or calculations of equilibrium for that system. Figure 4.11 repeats some of the examples used in this chapter but with the system boundary included for each one.

Gravity is always deemed to cross the boundary and enter the system (Figure 4.12). Equilibrium can only apply to a defined system. At the start of any calculations of equilibrium it is therefore important to state what system we are about to consider.

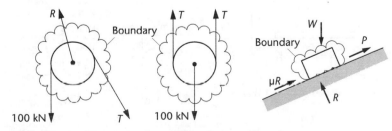

Figure 4.11   How to draw a boundary around systems.

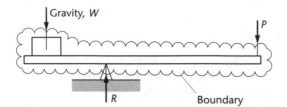

Figure 4.12 Gravity can always cross the system boundary.

➤ 4.4 Free body diagrams

In more complex problems, such as shown in Figure 4.13(a), it is useful to separate the whole into **parts** or **members** (Figure 4.13(b)). The term *component* may be used but must not be confused with *component of force* used in Section 3.3 and elsewhere, and *vector component* used in Section 2.3.

Each part is given a clear boundary, and all forces crossing that boundary constitute the system, as shown in Figure 4.13(b). Note that some of the forces transfer between adjacent systems, e.g. the force T_1 is transferred by the cable. When forces transfer in this manner they have the following properties:

➤ Equal in magnitude
➤ Opposite in direction
➤ Collinear

They fulfil the requirements of Section 3.8 and they obey Newton's third law.

A diagram like this showing part of a pulley arrangement, structural framework, etc., depicting *all* the forces of that system, is known as a **free body diagram**. Only the force system shown on such a free body diagram can be in equilibrium, and for this purpose what lies inside the system boundary becomes irrelevant (Figure 4.14).

A free body diagram may show one part (as already described), or many parts; it may even show a subdivided part (Figure 4.15). The choice of the system boundary in each case is arbitrary and only experience will allow systems to be selected profitably. In Figure 4.15(c) the difficulty of defining the forces entering the system at the broken

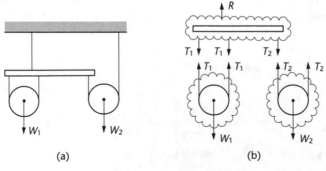

Figure 4.13 Complex problems may be separated into parts.

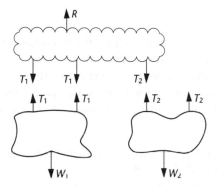

Figure 4.14 A free body diagram of the arrangement in Figure 4.13.

end arises because they are really internal forces (see Section 4.6). Each force system shown in the three parts of Figure 4.15 has a completely different set of forces, but each must be in equilibrium.

4.5 External forces

External forces applied to a structure are generally produced by gravity, wind, reactions, etc., and are resisted by internal forces generated within the structure. This statement is strictly true only when considering the whole of a structure. It should be modified to include the concepts of force systems (Section 4.3) and free body diagrams (Section 4.4). Hence, for any force system, the external forces are those crossing the boundary and entering the system. The internal forces are those contained within the boundary. So in Figure 4.15(a) forces R, T_1 and T_2 are all external forces, but in Figure 4.15(b) the forces R, W_1 and T_2 are external forces whereas the cable force T_1 is an internal force.

External forces commonly arise from gravity, reactions, wind pressure and suction, and water, earth or gas pressures; some are shown in Figure 4.16. Dynamic loads on structures are also external forces and may arise from vehicles, cranes and other machinery as a result of acceleration or braking (Figure 4.17). The force systems in Figures 4.16 and 4.17 constitute the whole structure.

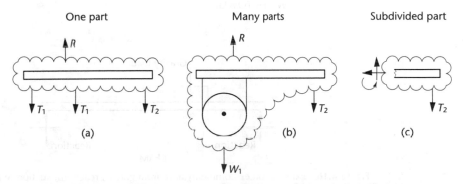

Figure 4.15 Free body diagrams may show one part, many parts or a subdivided part.

➤ 4.6 Internal forces

Internal forces are those totally contained within the boundary of the defined system. For a whole structure the internal forces may be cable tensions (T_1) as in Figure 4.15(b), or pin forces in a water-tower supporting frame (Figure 4.16), or bolt forces in a truss bridge girder. In addition internal forces may be contained within a part or member of a structure such as a beam or column. Four types of internal force are defined within such a member:

➤ **Axial** (direct) forces act along the axis of the member and may be either tension (pull) or compression (push) (Figure 4.18(a)).
➤ **Shear** forces act transverse to an axis but passing through that axis (Figure 4.18(b)).
➤ **Bending** acts as a transverse force with a lever arm, but passing through the axis of the member (Figure 4.18(c)).
➤ **Torsion** acts as a transverse force not intercepting an axis but at a lever arm out of the plane (Figure 4.18(d)).

All these internal forces have effects between the particles of the material from which

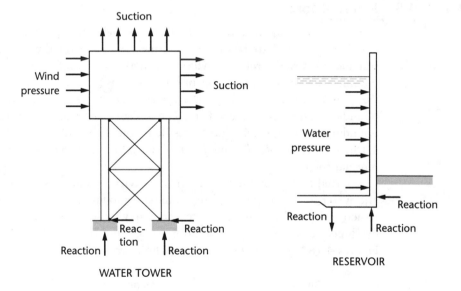

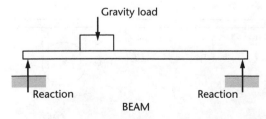

Figure 4.16 External forces commonly arise from gravity, reactions, suction, wind pressure and water pressure.

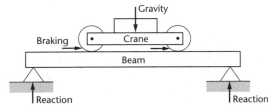

Figure 4.17 Dynamic loads are external forces.

the member is made, and give rise to stress, the subject of structural analysis and solid body mechanics. The simplest stress is due to an axial force P acting within a member of cross-sectional area A. The axial stress is given by P/A in units of $N\,m^{-2}$ or similar.

Every internal force may be resolved into a system of direct and shear stresses. Every stress system may be integrated back into one or more of the four internal forces.

4.7 Structural forms

Members in real structures have specific names and carry specific internal forces. Some of them are listed in Table 4.1.

Table 4.1 Structural forms

Structural form	Principal internal forces	Secondary internal forces
Columns (or struts)	Axial compression	Bending, shear
Ties (or hangers)	Axial tension	Bending, shear
Truss frames	Tension, compression	Bending, shear
Beams	Bending, shear	Axial
Edge beams	Bending, shear, torsion	Axial

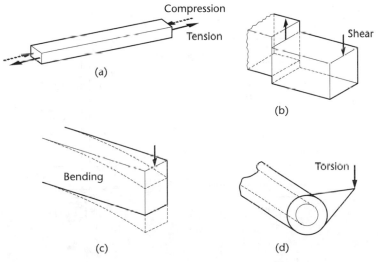

Figure 4.18 (a) Axial forces act along an axis. (b) Shear forces act transverse to an axis but passing through that axis. (c) Bending acts as a transverse force with a lever arm but passing through the axis of the member. (d) Torsion acts as a lever arm not intercepting an axis but at a lever arm out of the plane.

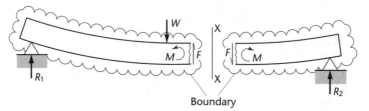

Figure 4.19 This beam has been cut at X–X to form two systems. The internal forces at X–X are a shear force and a bending moment; equal in sign but opposite in direction, these transferring forces will maintain equilibrium in each system.

Consider a beam which is bending under a load W, as shown in Figure 4.19, and imagine it to be cut at section X–X. The beam forms two force systems as shown and the internal forces at X–X are such as will maintain equilibrium in each system. The internal forces are a shear force and a bending moment. As noted in Section 4.4, the forces transferring at X–X are equal in magnitude and opposite in direction.

The radio tower shown in Figure 4.20(a) is loaded by external forces, gravity and wind; they are resisted by axial and shear forces and a bending moment. The motorway signboard in Figure 4.20(b) is similarly loaded, but the eccentricity of the load produces an extra force, a torsion.

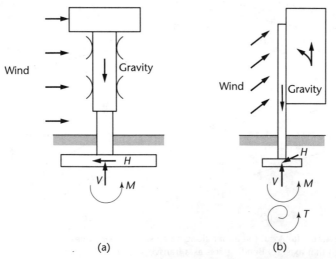

Figure 4.20 Two structures experiencing external forces: (a) a radio tower and (b) a motorway signboard. The eccentricity of the load in the signboard system produces a torsion.

Recap

■ **Normal reactions are perpendicular to contact surfaces.**

■ **Friction acts in the direction opposing motion.**

■ **Ropes and cables transmit tension only.**

■ **A force system must be defined before equilibrium can be considered.**

■ **Forces may be external or internal relative to the boundary of the force system.**

4.8 Problems

1. A cylinder of weight 4 kN rests on a plane inclined at 30° to the horizontal. A cable parallel to the plane and attached to the axis of the cylinder prevents movement. Find the cable force and the normal reaction.

2. If the cable in Problem 1 is adjusted to a horizontal position, find the new cable force and normal reaction.

3. In the pulley systems shown in Figure 4.21 find the cable tensions T and the axle reactions R; give the magnitude and direction for R.

4. In the pulley system shown in Figure 4.22 find the angles α_1 and α_2.

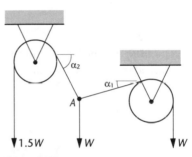

Figure 4.22

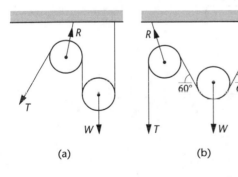

(a)   (b)

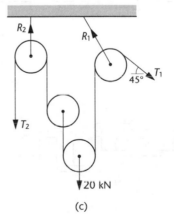

(c)

Figure 4.21

5. Consider the pulley system shown in Figure 4.23.

(a) Find the angle α for T_2 to be a minimum.

(b) Find the angle α for T_1 to be a minimum.

In case (b) what is the value of T_2?

In your answers to Problems 6 to 10, first define a boundary then a system of forces crossing that boundary. Finally apply the conditions of equilibrium to find the values of the forces required.

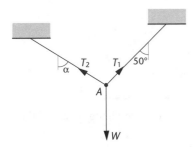

Figure 4.23

6. Find the reactions R_A and R_B for the beams shown in Figure 4.24. (The boundary encompasses the whole structure in each case.)

7. Find the forces in the struts *AC*, *AD* and *BE* of the structure shown in Figure 4.25. (For force direction in the struts see Section 3.8.)

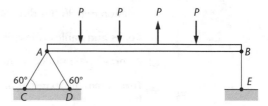

Figure 4.25

8. Find the force in the strut *BC* and the reaction at pin *A* in the structure shown in Figure 4.26.

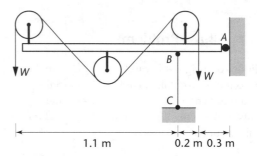

Figure 4.26

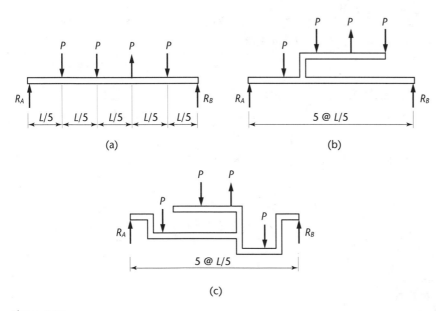

Figure 4.24

9. Find the reactions R_A and R_B supporting the wedge *ABC* (zero weight) shown in Figure 4.27. In (a) the boundary encompasses the whole structure. In (b) draw one boundary round the cylinder and one boundary round the wedge only.

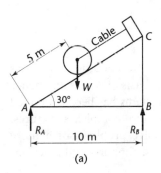
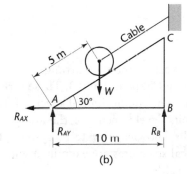

(a)   (b)

Figure 4.27

10. Find the tension in the string and the reaction transmitted at pin *A* of the structure shown in Figure 4.28. First use a boundary to encompass the whole structure and find R_B and R_C. Secondly use a boundary to encompass *AC* only.

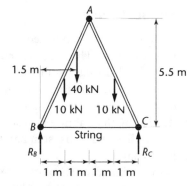

Figure 4.28

5 Frameworks

The concepts of force as vectors and their arrangement to give the condition known as equilibrium were set out in Chapters 2 and 3. These form the basis of statics and hence of all structural analysis. The equilibrium of specific force systems has been examined in Chapter 4 and permits development for the analysis of more complex structures. This chapter utilizes the knowledge of force systems to solve problems of a more complex nature. Practical structures are examined with loads, supports and reactions which would apply in real life.

5.1 Support types

Structures are supported on the ground, or on other structures known as substructures. The support may take the form of a hinge or a roller, or it may be fully fixed. Bridges sometimes include hinged supports, such as the Tyne Bridge, or roller (sliding) supports, as in many motorway bridges. Fully fixed supports are used in cantilever structures such as the signboard post in Figure 4.20(b). These supports are defined by their ability to resist forces and moments or to transfer them from one part to another. They are summarized in Figure 5.1, and the symbols shown are used in later chapters.

In a similar way, joints between members within a structure are defined by their ability to transmit force and moment. Sliding joints very rarely occur within structures, but pinned or fixed joints are common in frameworks.

5.2 Equilibrium of whole structure

Before considering the equilibrium of members, subframes or the free body diagrams (Section 4.4), it is useful to review the conditions of equilibrium (Section 3.7):

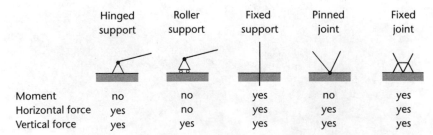

	Hinged support	Roller support	Fixed support	Pinned joint	Fixed joint
Moment	no	no	yes	no	yes
Horizontal force	yes	no	yes	yes	yes
Vertical force	yes	yes	yes	yes	yes

Figure 5.1 Conventional symbols for a variety of supports.

$$\sum F_x = 0$$

$$\sum F_y = 0$$

$$\sum M_o = 0$$

As a result of these conditions of equilibrium some special cases could be identified. Section 3.8 showed that when two forces act on a two-pinned member the force direction is along the line joining the pins. Section 3.9 showed that if three forces acting on a member are in equilibrium they meet at a common point or are parallel. In the examples that follow, simple structures are analysed by using the conditions of equilibrium and these two principles derived from them.

Example 5.1: ladder

Find the reactions at the ground and wall for the ladder shown in Figure 5.2 if there is (a) no friction at the wall and (b) friction of coefficient μ at the wall.

Solution In Figure 5.2(a) the three forces acting on the ladder are in equilibrium, so they meet at one point as shown. The force system includes the whole ladder (which is considered to be weightless) and the forces may be represented by the vector diagram shown. Hence

$R_1 = W/\tan \beta$ horizontal

$R_2 = W/\cos \beta$ at angle β with horizontal

Note that β is not the ladder angle α but the angle to the point where the forces meet. This may be found from the geometry of the diagram as follows:

$y = x \tan \alpha$

$\tan \beta = 2y/x$

$\quad = 2 \tan \alpha$

If friction equal to μH is present at the wall, the reaction at the wall is inclined to the horizontal as shown in Figure 5.2(b). The forces now meet at a different point, and the vector diagram changes accordingly.

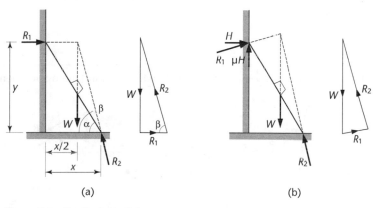

(a)   (b)

Figure 5.2   See Example 5.1.

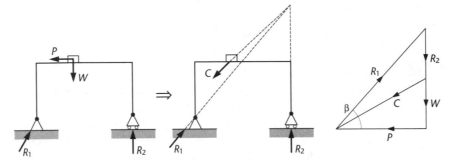

Figure 5.3 See Example 5.2.

➤ **Example 5.2: portal crane**

Find the reactions for the portal crane shown in Figure 5.3. The crab which moves along the top beam is acted on by gravity W and by a dynamic (braking) force P.

Solution The crane is acted on by four forces, so the number of forces must be reduced in order to use the three force principle. This can be achieved by the vector addition of the crane forces P and W to give force C. The force at the roller support is vertical (see Section 5.1), so the three forces form the system for the whole crane and they meet at the point shown. The vector diagram can be used to show the addition of P and W as well as the equilibrium of the three forces. From the vector diagram,

$$R_1 = P/\cos \beta$$

$$R_2 = P \tan \beta - W \text{ which is negative (i.e. downwards) for the vector diagram to close}$$

As before, the angle β is determined by the geometry of the crane.

➤ **5.3 Equilibrium of members**

Examples 5.1 and 5.2 considered the whole structure and applied the principles of equilibrium. For more complex structures it is necessary to consider individual members of the structure, or groups of members (subframes), when applying these principles.

➤ **Example 5.3: three-pinned arch**

Find the reactions for the three-pinned arch shown in Figure 5.4 when acted on by a gravity force W.

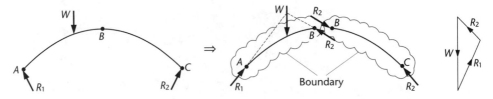

Figure 5.4 See Example 5.3.

Solution   The reactions are two vectors, with neither magnitude nor direction known in either case. This means there are four unknowns, and three equations of equilibrium for the whole arch cannot produce a solution directly. It is useful therefore to define two force systems with boundaries as shown. The part BC forms a two-force system and the direction of reaction R_2 is on the line joining the pins. This is a case where the line is not the axis of the member.

The part AB is acted on by three forces, one of which is the same force R_2, transferred at B with equal magnitude, collinear, but of opposite direction (see Section 4.4). These three forces meet at a point, so the vector diagram may be drawn and values for R_1 and R_2 may be found.

Example 5.4: hoist

The hoist shown in Figure 5.5 has a structure comprising a jib ABC and an arm BD. If a 20 kN load is supported at C find the reactions at A and D, and draw free body diagrams for both members.

Solution   As in the previous example, the reactions at A and D are vectors with a total of four unknowns. It is useful to define two force systems as shown. The free body diagrams for each member may be drawn, and the equilibrium conditions applied to each. The conditions of equilibrium (Section 3.5) are often expressed in terms of horizontal and vertical directions instead of x and y, that is

$$\sum H = 0, \quad \sum V = 0, \quad \sum M = 0$$

Member BD is a two-force system and the forces F_{BD} are as indicated. The jib ABC is acted on by three forces which meet at one point and a vector diagram may be drawn to find the values of F_{BD} and R_A. Alternatively, the hoist may be analysed by considering the equilibrium of each part, hence finding all the forces.

Equilibrium of jib ABC

$$\sum V = 0 \qquad\qquad V_A - 20 = 0$$
$$V_A = 20 \text{ kN}$$

$$\sum M_A = 0 \qquad 20 \times 10 - F_{BD} \times 3 = 0$$
$$F_{BD} = 66.7 \text{ kN}$$

$$\sum H = 0 \qquad\qquad H_A - 66.7 = 0$$
$$H_A = 66.7 \text{ kN}$$

The vector addition of V_A and H_A gives

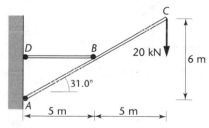

Figure 5.5   See Example 5.4.

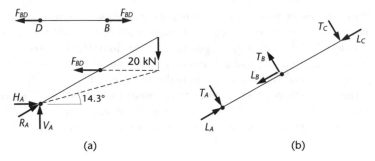

(a)   (b)

Figure 5.6 See Example 5.4.

$$RA = \sqrt{(20^2 + 66.7^2)}$$

$$= \textbf{69.6 kN at 16.7}° \textbf{ to the horizontal}$$

The free body diagrams for each member may be completed with the force vectors as shown in Figure 5.6(a). The diagram for *ABC* is correct, but in engineering it is more convenient to give each force in terms of longitudinal (*L*) and transverse (*T*) components (Section 3.3) as shown in Figure 5.6(b):

$$L_A = 69.6 \cos 14.3° = 67.5 \text{ kN}$$

$$T_A = 69.6 \sin 14.3° = 17.2 \text{ kN}$$

$$L_B = 66.7 \cos 31.0° = 57.2 \text{ kN}$$

$$T_B = 66.7 \sin 31.0° = 34.3 \text{ kN}$$

$$L_C = 20 \cos 59.0° \quad = 10.3 \text{ kN}$$

$$T_C = 20 \sin 59.0° \quad = 17.1 \text{ kN}$$

➤ Example 5.5: roof truss

The roof truss shown in Figure 5.7 supports loads of 5 kN, 10 kN and 5 kN perpendicular to the rafter. Find the reactions at *A* and *B*.

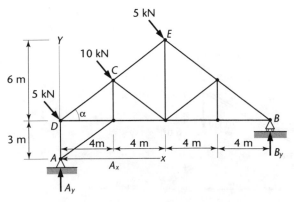

Figure 5.7 See Example 5.5.

Solution   A possible method of solution is to reduce the number of forces to three by combining the rafter forces to one of 20 kN acting at the centre of force C. Then the three force principle may be used as before in conjunction with a vector diagram.

But this example uses components of the forces as in Example 3.5. Each component is listed together with its perpendicular distance from some reference point, in this case A. The roof truss angle must be calculated.

$$\tan \alpha = 6/8 \qquad \text{hence } \alpha = 36.9°$$

Hence the loads at C, D and E act at a polar angle of 306.9°. The requirements of equilibrium may be set out in tabular form (Table 5.1) as in Section 3.7.

Table 5.1   Roof truss equilibrium

Position	Force (kN)	F_x (kN)	F_y (kN)	r_y (m)	r_x (m)	M_A (kN m)
A	R_A	A_x		0		0
			A_y		0	0
B	R_B	0		3		0
			B_y		16	$-16B_y$
C	10	10 cos 306.9°		6		36
			10 sin 306.9°		4	32
D	5	5 cos 306.9°		3		9
			5 sin 306.9°		0	0
E	5	5 cos 306.9°		9		27
			5 sin 306.9°		8	32

Note that M_A values must use a sign convention: clockwise positive, anticlockwise negative. The equilibrium of the whole truss implies

$$\sum M_A = 0 \qquad -16B_Y + 136 = 0$$

$$B_y = 8.5 \text{ kN}$$

$$\sum F_x = 0 \qquad A_x + 20 \cos 306.9° = 0$$

$$A_x = -12 \text{ kN}$$

$$\sum F_y = 0 \quad A_Y + B_Y + 20 \sin 306.9° = 0$$

$$A_y = 7.5 \text{ kN}$$

$$R_A = \sqrt{(12^2 + 7.5^2)}$$

$$= 14.2 \text{ kN}$$

Example 5.6: excavator

A mechanical excavator has the structure shown in Figure 5.8. The jib BCD and the arm ECF are pinned together at C. Hydraulic rams at D and F operate the machine, and for the position shown they maintain equilibrium. A load of 50 kN acts at E perpendicular to ECF.

(a) Find the horizontal and vertical components of the reactions at A and B (due to the 50 kN load only).

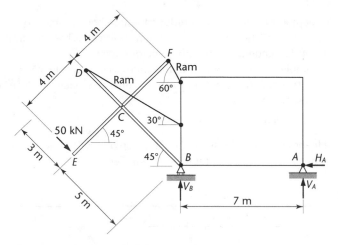

Figure 5.8 See Example 5.6.

(b) Find the forces in the rams at *D* and *F*.
(c) Draw the free body diagrams for the jib *BCD* and the arm *ECF*.

Solution   **Equilibrium of the whole excavator**

$$\sum M_B = 0 \qquad\qquad 50 \times 4 + V_A \times 7 = 0 \qquad V_A = -28.6 \text{ kN}$$

$$\sum H = 0 \qquad\qquad 50 \cos 45° - H_A = 0 \qquad H_A = 35.4 \text{ kN}$$

$$\sum V = 0 \qquad 50 \cos 45° + 28.6 - V_B = 0 \qquad V_B = 64.0 \text{ kN}$$

Equilibrium of arm *ECF*

Figure 5.9(a) shows the free body diagram for the arm *ECF*. As described in Example 5.4, and in Section 3.3, longitudinal and transverse components of each force are the most convenient vector description. Hence the conditions of equilibrium become

$$\sum L = 0, \qquad\qquad \sum T = 0, \quad \sum M = 0$$
$$\sum M_C = 0 \qquad 50 \times 4 - T_F \times 4 = 0 \qquad T_F = 50 \text{ kN}$$

The hydraulic ram at *F* is a two-force system, so the force P_F in the ram has a line of action along the ram axis.

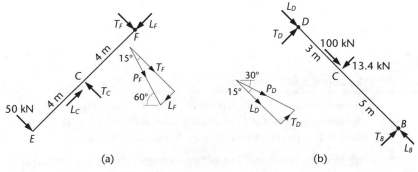

(a)   (b)

Figure 5.9 See Example 5.6.

$$L_F = 50 \tan 15° \qquad L_F = 13.4 \text{ kN}$$
$$P_F = 50/\sin 15° \qquad P_F = 51.8 \text{ kN}$$

$$\sum T = 0 \qquad 50 + 50 - T_C = 0 \qquad T_C = 100 \text{ kN}$$
$$\sum L = 0 \qquad 13.4 - L_C = 0 \qquad L_C = 13.4 \text{ kN}$$

Equilibrium of jib BCD

Figure 5.9(b) shows the free body diagram for the jib *BCD*. Note carefully the transfer of the forces T_C (100 kN) and L_C (13.4 kN) from member *ECF* to member *BCD*. The forces transferred must be equal in magnitude, opposite in direction and collinear (see Section 3.8).

$$\sum M_B = 0 \qquad 13.4 \times 5 - T_D \times 8 = 0 \qquad T_D = 8.4 \text{ kN}$$
$$\sum T = 0 \qquad 13.4 - 8.4 - T_B = 0 \qquad T_B = 5.0 \text{ kN}$$

The hydraulic ram at *D* is a two-force system and the direction of the force P_D is known as before.

$$L_D = 8.4/\tan 15° \qquad L_D = 31.4 \text{ kN}$$
$$P_D = 8.4/\cos 15° \qquad P_D = 32.4 \text{ kN}$$
$$\sum L = 0 \qquad 31.4 + 100 - L_B = 0 \qquad L_B = 131.4 \text{ kN}$$

5.4   Overturning

Overturning is a possibility in many engineering structures and machinery. Examples are mobile plant, such as cranes and excavators (Figure 5.8), and certain structures such as bridges. All structures must also be assessed for stability, which includes overturning.

Mobile plant, such as a crane, might overturn about a particular point, such as the front wheels. This point is then the fulcrum of overturning. At the time when overturning is about to happen, the reaction on the other (rear) wheels has become zero and all the reaction is concentrated at the fulcrum. At this time the overturning moment approximately balances the righting moment, which is preventing overturning. In practice the overturning moment is provided by a load (such as the 50 kN in Figure 5.8), and the righting moment is given by the vehicle weight (not shown in Figure 5.8). It is not safe to operate machines when the overturning and righting moments just balance; this is because overturning will be imminent. It is therefore usual to introduce a factor of safety. If the righting moment were twice the overturning moment, this would provide a safety factor of 2 against overturning.

Example 5.7: cleaning cradle

A window-cleaning cradle support is structured as shown in Figure 5.10 and loaded at *C* by the cradle and operatives. If a wheel at *A* can only provide upward (not downward) reaction, find what counterbalance weights must be provided at *A* to give a safety factor of 2 against overturning.

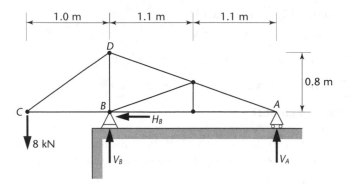

Figure 5.10 See Example 5.7.

Solution **Equilibrium of whole structure**

The reaction at A is taken as zero on overturning. If weights W are placed at A, as shown in Figure 5.11(a), then

Overturning moment	$M_o = 8 \times 1.0 = 8$ kN m
Righting moment	$M_R = W \times 2.2 = 2.2W$ kN m
For safety	$M_R = 2M_o$
	$2.2W = 2 \times 8$
	$W = \mathbf{7.27}$ **kN**

If instead of weights a tieback is provided at point D (Figure 5.11(b)), find the necessary strength of the tie.

Equilibrium of whole structure

As before	$M_o = 8$ kN m
Righting moment	$M_R = H \times 0.8 = 0.8$ kN m
For safety	$M_R = 2M_o$
	$0.8H = 2 \times 8$
	$H = \mathbf{20}$ **kN**

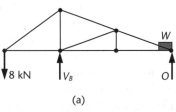

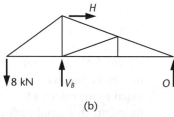

(a) (b)

Figure 5.11 See Example 5.7.

> ## Recap
>
> ■ **Reactions may be found by considering equilibrium of the whole framework.**
>
> ■ **Forces between members and within members may be found by equilibrium of individual members or groups of members.**
>
> ■ **Overturning acts about a fulcrum and includes a factor of safety.**

➤ 5.5 Problems

1. Determine graphically and analytically the forces which must be applied in directions *AE* and *AF* so the joint at *A* shown in Figure 5.12 shall be in equilibrium.

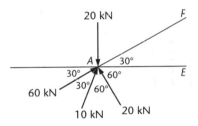

Figure 5.12

2. Figure 5.13 shows the arrangement at the top of a 200 kN jib crane. Pulley 2 is used to permit change in the length of the tie and alteration of the jib angle. Draw force systems for each of the members (hook, pulley 1, pulley 2, jib) and determine the tension in load cable (T_1), tie cable (T_2), force in the jib (F_1) and pull (P).

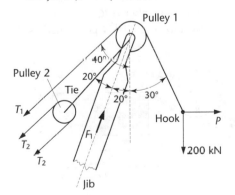

Figure 5.13

3. By drawing the force system for the whole structure, find the force in the cable and the support reaction for the hoist shown in Figure 5.14.

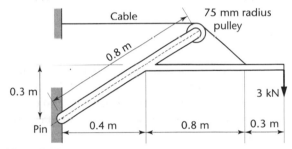

Figure 5.14

4. A column *ABC* of a building has a cantilever truss attached to it at *A* and *B* (Figure 5.15). The column receives horizontal support from the main roof at *A* and is hinged at the base *C*. Determine the force F_H and the reaction at *C*

(a) analytically and

(b) using the three force condition.

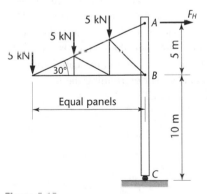

Figure 5.15

5. The left-hand portion of a spandrel-braced arch is shown in Figure 5.16. The arch is hinged at *A* and is supported at *B* by a force acting in the direction *BC*. Determine the value of the reactions at *A* and *B* (a) analytically and (b) using the three force condition.

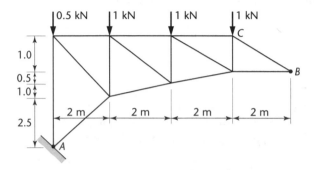

Figure 5.16

6. Find the horizontal and vertical components of the reaction at *A* and the force in the jack *BC* for the inspection platform shown in Figure 5.17; (a) use the three force condition and the force diagram, (b) use analysis.

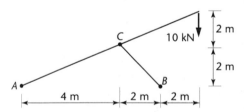

Figure 5.17

7. The structure shown in Figure 5.18 consists of two separate beams *ABC* and *CD*, pinned together at *C* over one support. Member *EB* is connected by a rigid joint at *B* to *ABC*, and there is a flexible cable connecting *DE*. Find the force in the cable and the reactions at *A* and *C* by determining the force systems for *ABCE* and *CD*.

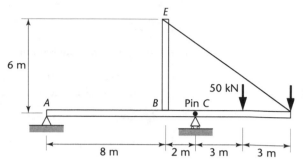

Figure 5.18

8. An inspection gantry *ABCD* is subjected to vertical loads as shown in Figure 5.19. Use a graphical means to obtain the magnitude and direction of the reactions at *A* and *D*. An additional horizontal load of 1 kN is applied at *B* in the direction *BC*; indicate its effects on your diagrams. New magnitudes of the reactions need not be given.

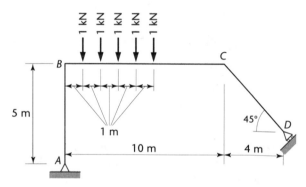

Figure 5.19

9. The basic structure of a bulldozer is shown in Figure 5.20. Hydraulic rams *BE* and *CF* are attached to the frame member *ABCD*. The blade *AE* is acted on by a force of 40 kN positioned as

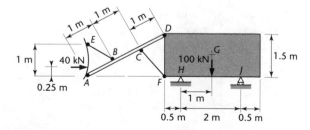

Figure 5.20

shown. The vehicle weighs 100 kN, which may be considered to act at G.

(a) Find the reactions at H and J.

(b) Find the forces in the hydraulic rams BE and CF.

(c) Draw the free body diagram for member ABCD.

10. For a pin-ended strut, having forces applied at the pins only, show that the direction of the forces is on the line joining the pins. The gantry crane shown in Figure 5.21 is pinned at A, B and C, and supports crabs at E and F. Each crab is loaded vertically and horizontally as shown. Find the magnitude and direction of the reactions at A and C.

11. A mechanical excavator weighing 60 kN has the structure shown in Figure 5.22. The members CDE and DFGH are pinned together at D. Hydraulic rams EF and GJ are pinned to the main members and to the cab. A force of 35 kN acts at an angle of 45° to the horizontal at C.

(a) Calculate the factor of safety against overturning about A.

(b) Calculate the reactions at A and B.

(c) Draw the free body diagram for member CDE.

(d) Calculate the force in ram EF.

(e) Calculate the force in ram GJ.

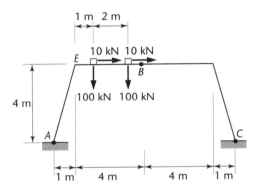

Figure 5.21

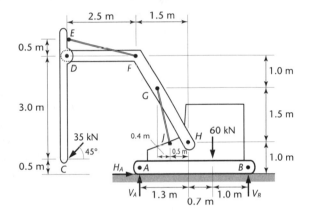

Figure 5.22

6 Plane frames

The principles of equilibrium have been examined in previous chapters, where they were applied to single points, to larger members and to complex groups known as structures. In this chapter the principles of equilibrium will be applied to a specific group of structures known as **statically determinate** frameworks. Formed of simple bar members, lying in one plane and pinned together at joints, statically determinate frameworks are often known as two-dimensional pin-jointed frames. The methods used in Chapters 4 and 5 can be applied to these frames with equal success: we may apply equilibrium to individual parts, subframes or complete structures, in order to obtain solutions.

6.1 Members and frames

In the design and analysis of plane frames, members are considered as pinned at both ends. Hence the force at each pin, applied to the member, acts along a line joining the pins (Figure 6.1). This rule was examined and proved in Section 3.8 and applies regardless of the shape of the member (Figure 6.2). In practical cases, such as a roof truss, forces may be applied at other points besides the pins, and this invalidates the rule. However, design in practice assumes the pin-jointed rule, and makes a modification to allow for the effects of bending.

In the usual case of straight members, pinned at both ends, the internal forces are axial and may be either tension (pull) or compression (push).

The simplest plane frame is a triangle formed by three members pinned together as shown in Figure 6.3. More complex plane frames can be arranged by adding members in pairs as shown in Figure 6.4. Note that each extra pair of members locates one extra joint in the plane.

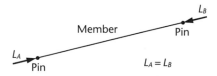

Figure 6.1 The force acts along a line joining the pins.

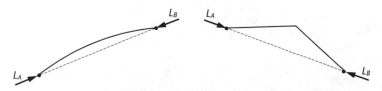

Figure 6.2 Regardless of the shape of the member, the force acts along a line joining the pins.

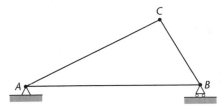

Figure 6.3 The simplest plane frame is a triangle of members.

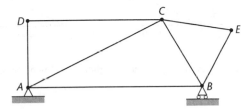

Figure 6.4 Each extra pair of members locates one extra joint in the plane.

6.2 Equilibrium and reactions

As set out in Sections 4.3 and 4.4, force systems define that particular set of forces which are in equilibrium. The system may be made up of a single item, many items, or all of the structure. For plane frames free body diagrams may be used for each member, or a specific joint (or pin), or for the whole frame (Figure 6.5). Use of the free body diagram for a group of members will be examined in Section 6.5.

Note that in Figure 6.5 each force system is in equilibrium but sustains a different set of forces. The action of a typical force P on the whole frame requires three reactions (Figure 6.6) to maintain equilibrium. These reactions together with the force P satisfy the three equations of equilibrium (Section 3.7). The reactions may be viewed as preventing movement in the vertical and horizontal directions as well as preventing rotational movement (Figure 6.7).

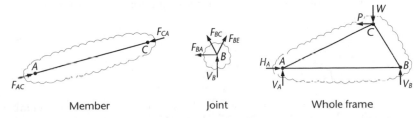

Member   Joint   Whole frame

Figure 6.5 Free body diagrams may be drawn for a member, a joint or a whole frame.

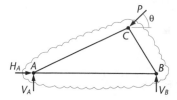

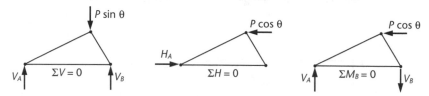

Figure 6.6 The action of a typical force *P* requires three reactions to maintain equilibrium.

Figure 6.7 The reactions can be viewed as preventing vertical movement, horizontal movement and rotational movement.

➤ 6.3   Simple stiffness

Practical plane frames are used as roof trusses, bridge girders, towers, masts and wind bracing. They range in size from a few members to many hundreds of members. The arrangement of members may be 'simple' or 'complex'. As far as analysis is concerned, simple frames are those which can be solved by application of the principles of statics (equilibrium) alone. They are therefore 'statically determinate', and hence come within the scope of this chapter.

The complex frames cannot be solved by application of the principles of equilibrium alone. Besides equilibrium their solution requires some knowledge of the geometric **fit**, i.e. principles of compatibility. These frames are **statically indeterminate** (or hyperstatic, or redundant) and are outside the scope of this book.

As shown in Figures 6.3 and 6.4, the number of joints in a statically determinate frame is linked to the number of members:

A simple relationship is immediately apparent:

Joints (*j*)	3	4	5	6	. . .
Members (*m*)	3	5	7	9	. . .

$$m = 2j - 3$$

Considerations of the equations of equilibrium together with the number of unknown quantities gives this relationship in greater detail. The number of equations available in a simple frame is *2j*, since two equations of equilibrium may be identified at each joint.

$$\sum F_x = 0$$

$$\sum F_y = 0$$

The number of unknowns is the sum of the number of member forces (*m*) and the number of reactions (*r*). A solution by statics alone is only possible when the number of equations and the number of unknowns are equal, i.e.

$$2j = m + r$$

$$m = 2j - r$$

For the general loading case the number of reactions is a minimum of three (Section 6.2), but as shown in Figure 6.8, more reactions are possible without violating this test for simple stiffness.

Examples of different kinds of frame, with different layouts of members, joints and reactions, are shown in Figure 6.8. Figure 6.8(a) and (b) satisfy the test for simple stiffness, so the frames are statically determinate. But Figure 6.8(c) contains three more unknowns $(m + r)$ than the number of equations $(2j)$. This frame is therefore described as three times statically indeterminate.

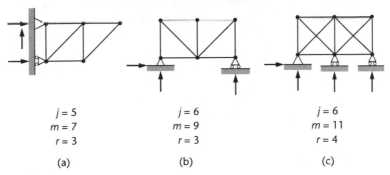

$j = 5$
$m = 7$
$r = 3$

(a)

$j = 6$
$m = 9$
$r = 3$

(b)

$j = 6$
$m = 11$
$r = 4$

(c)

Figure 6.8 Using the equation $2j = m + r$, it is possible to show that (a) is statically determinate, (b) is statically determinate and (c) is statically indeterminate.

6.4 Frame analysis

All forces acting on the frame or at specific joints are vectors, so the principles of equilibrium will involve drawing force diagrams or using calculation as described in Section 3.5. Figure 6.9 shows a typical joint in a simple frame and the vector diagram showing equilibrium of the forces.

Solution of the whole frame may be carried out by proceeding from joint to joint, transferring member forces as shown in Figure 6.10. When transferring member forces to the next joint, the force direction is reversed to satisfy the condition of equilibrium of the two-pinned member (see Section 3.8). Analysis in this manner, using calculation at each joint, is known as solution by the method of joint resolution.

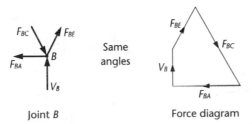

Joint B

Force diagram

Figure 6.9 A typical joint in a simple frame accompanied by its vector diagram.

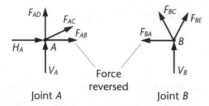

Figure 6.10 When transferring member forces to the next joint, the force direction is reversed to satisfy the condition of equilibrium in the two-pinned member.

Each joint may be solved by using a force diagram (Figure 6.9). This **method of force diagrams** may be extended so the diagrams for individual joints are joined together to form one large force diagram. This arrangement is simplified by using Bow's notation, the principal graphical method for solving pin-jointed frames. Details of this method lie outside the scope of this book.

A very useful method of solution for simple frames is the **method of sections**. Examined in detail in Section 6.5, it uses free body diagrams made up of several joints and members. It lies between considering equilibrium of single joints and considering equilibrium of the whole frame (Figure 6.5). The force systems are intermediate between the two, and it is therefore similar to the subgroup arrangement of Figure 4.15(b). It is particularly useful when only a few member forces are required.

➤ **Example 6.1**

Find the reactions and forces in the members of the frame shown in Figure 6.11 as a result of a vertical load W at C.

Solution **Equilibrium of whole frame** (Figure 6.11)

$$\sum H = 0$$

hence $H_B = 0$

$$\sum M_A = 0$$

$$W \times L/2 - V_B \times L = 0 \qquad V_B = W/2$$

$$\sum V = 0$$

$$V_A + W/2 - W = 0 \qquad V_A = W/2$$

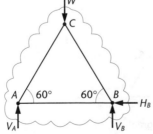

Figure 6.11 See Example 6.1.

Equilibrium of joint *A* (Figure 6.12)

$$\sum V = 0$$

$$W/2 + F_{AC}\cos 30° = 0 \qquad F_{AC} = -W/\sqrt{3}$$

The negative sign for F_{AC} indicates it is in the opposite direction from that shown at joint *A* (Figure 6.12). This is corrected before continuing to joint *C*.

$$\sum H = 0$$

$$F_{AB} - (W/\sqrt{3})\cos 60° = 0 \qquad F_{AB} = W/2\sqrt{3}$$

Equilibrium of joint *C* (Figure 6.12)

$$\sum H = 0$$

$$(W/\sqrt{3})\cos 60° + F_{BC}\cos 60° = 0 \qquad F_{BC} = -W/\sqrt{3}$$

$$\sum V = 0$$

$$(W/\sqrt{3})\cos 30° + (W/\sqrt{3})\cos 30° - W = 0$$

which is a check calculation

Equilibrium of joint *B*

$$\sum H = 0 \qquad \text{(check)}$$

$$\sum V = 0 \qquad \text{(check)}$$

This example was solved without reference to the dimensions of the frame, and depended on the relative angles of the members alone.

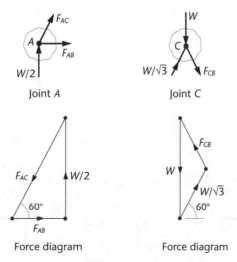

Joint *A*   Joint *C*

Force diagram   Force diagram

Figure 6.12   See Example 6.1.

➤ **Example 6.2**

Find the forces in the members of the frame shown in Figure 6.13 as a result of two loads W at C and D.

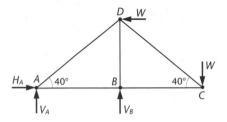

Figure 6.13 See Example 6.2.

Solution **Equilibrium of joint C** (Figure 6.14)

$$\sum V = 0$$

$$F_{CD} \cos 50° - W = 0 \qquad F_{CD} = 1.556W$$

$$\sum H = 0$$

$$1.556W \cos 40° + F_{CB} = 0 \qquad F_{CB} = -1.192W$$

The negative sign for F_{CB} gives an opposite direction as in Example 6.1.

Equilibrium of joint D (Figure 6.14)

$$\sum H = 0$$

$$1.556W \cos 40° - W - F_{DA} \cos 40° = 0 \qquad F_{DA} = 0.251W$$

$$\sum V = 0$$

$$1.556W \cos 50° + 0.251W \cos 50° + F_{DB} = 0 \qquad F_{DB} = -1.162W$$

Equilibrium of joint B (Figure 6.14)

$$\sum V = 0$$

$$V_B = 1.162W$$

$$\sum H = 0$$

$$F_{AB} = -1.192W$$

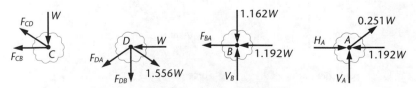

Figure 6.14 See Example 6.2.

Equilibrium of joint *A* (Figure 6.14)

$$\sum H = 0$$

$$H_A - 1.192W + 0.251W \cos 40° = 0 \qquad H_A = W$$

$$\sum V = 0$$

$$V_A + 0.251W \cos 50° = 0 \qquad\qquad V_A = -0.161W$$

Example 6.3

Find the forces in the members of the pin-jointed plane frame shown in Figure 6.15.

Solution **Equilibrium of the whole structure** (Figure 6.15)

$$\sum M_E = 0$$

$$6V_A - 4 \times 5 + 2 \times 25 = 0 \qquad\qquad V_A = -5 \text{ kN}$$

$$\sum V = 0$$

$$V_A + V_E - 10 - 25 = 0 \qquad\qquad V_E = 40 \text{ kN}$$

$$\sum H = 0$$

$$H_E - 5 = 0 \qquad\qquad H_E = 5 \text{ kN}$$

Equilibrium of joint *A* (Figure 6.16)

$$\sum V = 0$$

$$-5 + P_{AB} \times 4/5 = 0 \qquad\qquad P_{AB} = 6.25 \text{ kN}$$

$$\sum H = 0$$

$$P_{AB} \times 3/5 + P_{AF} = 0 \qquad\qquad P_{AF} = -3.75 \text{ kN}$$

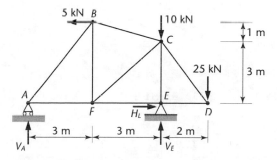

Figure 6.15 See Example 6.3.

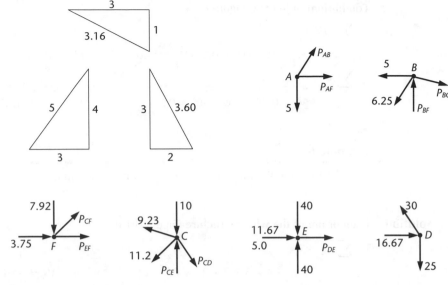

Figure 6.16 See Example 6.3.

Equilibrium of joint *B*

$$\sum H = 0$$

$$-5 - 6.25 \times 3/5 + P_{BC} \times 3/3.16 = 0 \qquad P_{BC} = 9.23 \text{ kN}$$

$$\sum V = 0$$

$$-6.25 \times 4/5 + 9.23 \times 1/3.16 + P_{BF} = 0 \qquad P_{BF} = 7.92 \text{ kN}$$

Equilibrium of joint *F*

$$\sum V = 0$$

$$-7.92 + P_{CF} \times 1/\sqrt{2} = 0 \qquad\qquad P_{CF} = 11.2 \text{ kN}$$

$$\sum H = 0$$

$$3.75 + P_{EF} + 11.2 \times 1/\sqrt{2} = 0 \qquad\qquad P_{EF} = -11.67 \text{ kN}$$

Equilibrium of joint *C*

$$\sum H = 0$$

$$-11.2 \times 1/\sqrt{2} - 9.23 \times 3/3.16 + P_{CD} \times 2/3.6 = 0 \qquad P_{CD} = 30 \text{ kN}$$

$$\sum V = 0$$

$$-10 - 30 \times 3/3.6 - 11.2 \times 1/\sqrt{2} + 9.23 \times 1/3.16$$
$$+ P_{CE} = 0 \qquad\qquad P_{CE} = 40 \text{ kN}$$

Equilibrium of joint E

$$\sum H = 0$$

$$11.67 + 5 + P_{DE} = 0 \qquad\qquad P_{DE} = -16.67 \text{ kN}$$

$$\sum V = 0$$

$$V_E - P_{CE} = 0 \qquad\qquad \text{(check)}$$

Equilibrium of joint D

$$\sum H = 0$$

$$P_{DE} - 30 \times 2/3.6 = 0 \qquad\qquad \text{(check)}$$

$$\sum V = 0$$

$$-25 + 30 \times 3/3.6 = 0 \qquad\qquad \text{(check)}$$

Member forces (kN)

AB	AF	BC	BF	CD	CE	CF	DE	EF
6.25	3.75	9.23	7.92	30.0	40.0	11.2	16.67	11.67
t	c	t	c	t	c	t	c	c

t = tension, c = compression

6.5   Method of sections

Section 4.4 showed how free body diagrams may be chosen to represent one part, many parts or the whole structure. Force systems could be defined for any one of these arrangements, and the principles of equilibrium could be applied. The method of sections uses free body diagrams of many parts (Figure 4.15(b)), with the force system chosen as convenient, usually with the aid of experience.

Figure 6.17(a) shows a simple frame in which only the forces in members AB, BC and CD are required. A force system may be defined as shown, the boundary of which cuts through the three members. These member forces now pass through the boundary and must be taken into account when considering the equilibrium of the system. The system could be defined alternatively by a cut line or 'section' as shown in Figure 6.17(b), hence the name method of sections.

Equilibrium of the force system may now be considered and will involve the forces shown in Figure 6.18. Any set (Section 3.7) of the three equilibrium conditions may now be used, such as

$$\sum F_x = 0$$

$$\sum F_y = 0$$

$$\sum M_o = 0$$

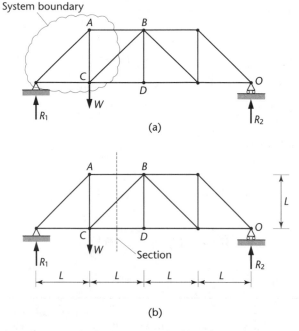

Figure 6.17 Method of sections: (a) a boundary is drawn on the free body diagram to intersect members *AB, BC* and *CD*; (b) the boundary is replaced by the conventional chain-dotted section line that intersects the same three members.

Hence, it is necessary to have no more than three unknown forces if a solution is to be obtained. Experience is valuable when drawing the system boundary to include only three unknown forces, the member forces required. The member forces will be tensions or compressions as described in Section 6.1.

➤ **Example 6.4**

Analyse the frame shown in Figure 6.17(a) by the method of sections to give the forces in members *AB, BC* and *CD* only.

Solution   **Equilibrium of whole structure**

$$\sum M_o = 0$$

$$R_1 \times 4L - W \times 3L = 0 \qquad\qquad R_1 = 3W/4$$

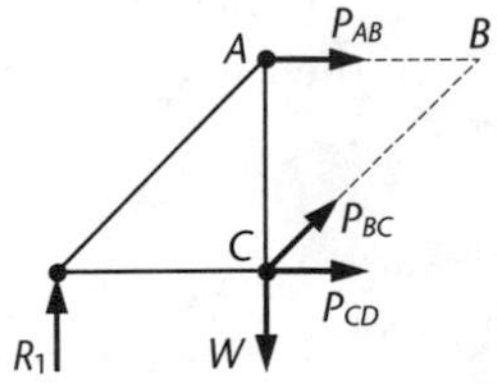

Figure 6.18 This section of the free body diagram in Figure 6.17 is an equilibrium system. The forces depicted here must satisfy the sets of equilibrium conditions in Table 3.1.

Equilibrium of force system (Figure 6.18)

$$\sum M_C = 0$$

$$P_{AB} \times L + R_1 \times L = 0 \qquad\qquad P_{AB} = -3W/4$$

$$\sum M_B = 0$$

$$R_1 \times 2L - W \times L - P_{CD} \times L = 0 \qquad P_{CD} = W/2$$

$$\sum F_y = 0$$

$$P_{BC} \cos 45° + R_1 - W = 0 \qquad\qquad P_{BC} = 0.35W$$

This example used the equilibrium conditions in Set 3 of Table 3.1:

$$\sum F_y = 0$$
$$\sum M_C = 0$$
$$\sum M_B = 0$$

▶ **Example 6.5**

Find the forces in members *AB*, *BC* and *CD* only for the pin-jointed plane frame shown in Figure 6.19.

Solution Consider a section at X–X, i.e. arrange a force system to enclose only the left-hand side of the frame (Figure 6.20).

$$\theta = \tan^{-1} 0.2/1.0 = 11.3°$$

$$BC = 0.5\sqrt{2} = 0.707 \text{ m}$$

$$CE = 0.707 \sin(45° - \theta) = 0.392 \text{ m}$$

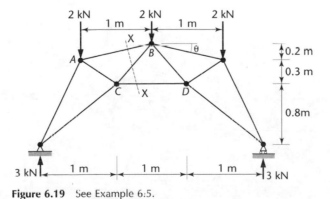

Figure 6.19 See Example 6.5.

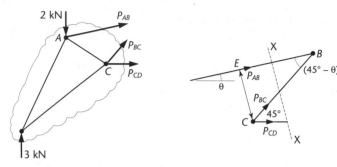

Figure 6.20 See Example 6.5.

Equilibrium of the left-hand force system means

$$\sum M_B = 0$$

$$3 \times 1.5 - 2 \times 1.0 - P_{CD} \times 0.5 = 0 \qquad P_{CD} = 5.0 \text{ kN}$$

$$\sum M_C = 0$$

$$3 \times 1.0 - 2 \times 0.5 + P_{AB} \times 0.392 = 0 \qquad P_{AB} = -5.10 \text{ kN}$$

$$\sum V = 0$$

$$3 - 2 - 5.1 \sin 11.3° + P_{BC} \times 1/\sqrt{2} = 0 \qquad P_{BC} = 0$$

Note that member forces may be zero. But this does not make the member redundant as defined in Section 6.3.

Recap

◼ **Forces in members in pin-jointed plane frames act along the line joining the pins.**

◼ **Forces may be found by considering the whole frame, part of a frame or each joint.**

◼ **Statically determinate frames satisfy the equation $m = 2j - r$.**

6.6 Problems

In answering each question, first define the part for which you are considering equilibrium. Draw the boundary and define the force system if necessary.

1. Find the reactions only of the pin-jointed plane frame shown in Figure 6.21.

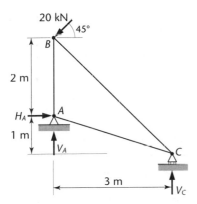

Figure 6.21

2. Find the reactions only of the pin-jointed plane frame shown in Figure 6.22.

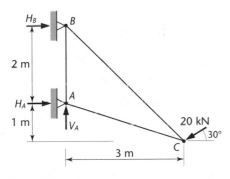

Figure 6.22

3. Find the reactions and the forces in the members of the plane frame shown in Figure 6.23.

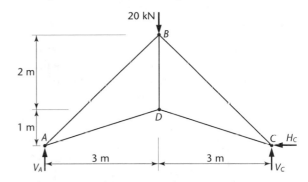

Figure 6.23

4. Find the reactions and the forces in the members of the plane frame shown in Figure 6.24.

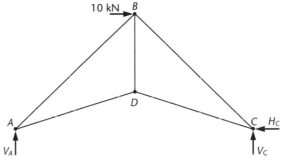

Figure 6.24

5. Find the reactions and the forces in the members of the plane frame shown in Figure 6.25.

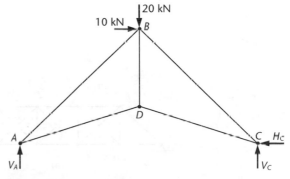

Figure 6.25

Note the connection between the answers to Problems 3, 4 and 5. The reactions and member forces in Problem 5 may be obtained from the addition of the same answers in Problems 3 and 4. This is an illustration of the principle of superposition.

6. Find the forces in the members *ab*, *bc* and *ac* of the pin-jointed plane frame shown in Figure 6.26. Use the method of sections and resolution of forces at joint *b*.

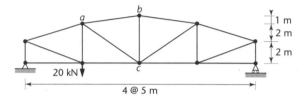

Figure 6.26

7. Show that the frame shown in Figure 6.27 is statically determinate. Find the force in the member *ab* using joint resolution, and verify your answer by the method of sections.

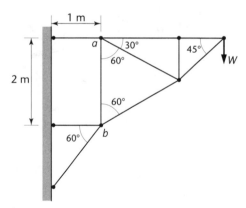

Figure 6.27

8. In the pin-jointed truss shown in Figure 6.28 all angles are either 45° or 90°. Using the method of sections, determine the forces in members *AB*, *BC* and *CD* when the truss is supported and loaded as shown. If negative reactions are not possible, what is the factor of safety against overturning? (For a discussion of overturning see Section 5.4.)

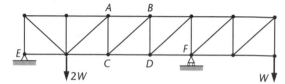

Figure 6.28

9. The pin-jointed frame shown in Figure 6.29 is supported by vertical reactions at *A* and *B*, and loaded as shown. Show that the frame is statically determinate. Find by the method of sections the forces in members *CD*, *DE* and *EC*. After finding the reactions, a solution is obtained by taking sections at the positions shown.

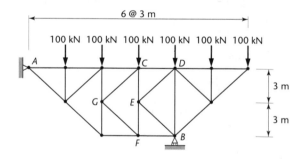

Figure 6.29

10. Find the degree of indeterminacy of the pin-jointed plane frames in Figure 6.30.

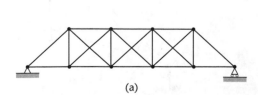

(a)

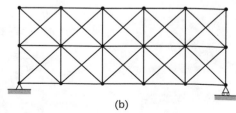

(b)

Figure 6.30

11. The pin-jointed frame shown in Figure 6.31 supports a load P. What is the maximum value of P if the stress in any member must not exceed $120\,\text{N}\,\text{mm}^{-2}$ in tension and $100\,\text{N}\,\text{mm}^{-2}$ in compression? All members have a cross-sectional area of $300\,\text{mm}^2$. (See Section 4.6 for simple stress concepts.)

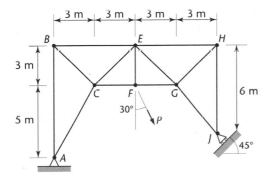

Figure 6.31

12. State whether or not the frames shown in Figures 6.32(a) and (b) are statically determinate. If not, what is the degree of statical indeterminacy? Find the forces in all the members of the pin-jointed frame shown in Figure 6.32(c).

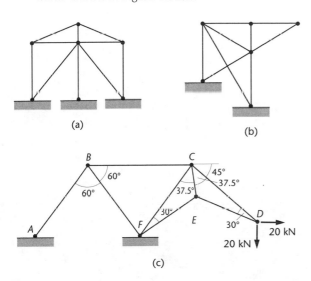

Figure 6.32

13. The trolley for a window-cleaning cradle is made from two frames, one of which is shown in Figure 6.33. The cradle is suspended at A and exerts a vertical load of $3\,\text{kN}$ on each frame. Wheels at B and C provide no resistance to uplift.

(a) Find the values of the counterweights W and $2W$ required to give a safety factor of 3 against overturning.

(b) Find the reactions on the parapet and roof, with the counterweights in place. Analyse the frame assuming pin-joints, and find the maximum tensile and compressive stresses if all members have a cross-sectional area of $120\,\text{mm}^2$.

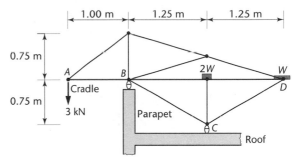

Figure 6.33

PART II Kinematics

Part II is devoted to kinematics, which studies the motions of bodies without reference to the forces causing them. This is the first of two parts on dynamics, the branch of mechanics concerned with the motions of bodies under the action of out-of-balance forces.

7 The motion of particles

As we saw in Chapter 3, when the forces acting on a body are balanced – their vector sum is zero – the body is in a state of equilibrium. It will therefore remain at rest or continue to move in a straight line with uniform velocity, unless it is acted upon by an unbalanced force. The study of problems such as these is encompassed by the branch of mechanics known as **statics**.

If a system of unbalanced forces acts on a body, it will be set in motion, or undergo a change in velocity if it is already moving. Parts II and III will be devoted to **dynamics**, the branch of mechanics concerned with the motions of bodies under the action of out-of-balance forces. The study of dynamics can itself be divided into two parts. The first is **kinematics**, which studies the motions of bodies without reference to the forces causing them. The second is **kinetics**, which studies the relationship between the motions of bodies and the forces that cause them.

Mechanics has traditionally been practised according to the precepts of Newton, now superseded by Einsteinian mechanics, based on the theory of relativity. Nonetheless, for most engineering and scientific purposes, Newtonian mechanics can be used effectively for solving problems. But it does not hold for problems encountered in astrophysics, where speeds may be much closer to the speed of light. The material in the following chapters will be set exclusively in the context of Newtonian mechanics.

A body is an object of finite size and mass. When acted on by a system of forces it may be expected to deform and perhaps to move. In many circumstances the deformation may be insignificant in comparison to the overall dimensions of the body, or to its changes in position as a whole. In this case it is convenient to consider it as a **rigid body**, for which the distance between any two points within the body is assumed to remain constant under the application of forces to the body.

Similarly, for many purposes in the study of dynamics the actual dimensions of the body are irrelevant to a description of the motion it is undergoing, or the action of the forces influencing that motion. In this instance it is convenient to consider the body to behave like a **particle**, defined as an object of finite mass but negligible dimensions. It is worth noting in passing, that **mass** is merely a measure of the amount of material in a body. Although it is responsible for the existence of gravitational attraction, it should not be confused with **weight**, which is the force exerted on a body by the gravitational pull of the earth.

It is customary, and logical, to look at the kinematics of bodies before proceeding to consider their kinetics. The basic principles can be most easily understood by starting with their application to particles. The rest of Part II will look at the motions of particles.

➤ 7.1 Types of motion and their definition

To study the motion of particles requires a description of their location in space at different times. This demands a reference point and a method of defining the position of the particle in relation to that reference point. The basic frame of reference in Newtonian mechanics, sometimes called the **primary inertial system**, is a rectangular set of axes assumed to have no translation or rotation in space. Any measurements made with respect to this fixed coordinate system are described as **absolute** quantities. In this book, Newton's laws can be deemed to hold in any frame of reference, either fixed or moving with constant velocity, in relation to the primary inertial system. We shall take our frames of reference as being fixed to the earth.

Three measurements are required to fix a point in three-dimensional space relative to some reference. Only two are needed in two-dimensional space. The measurements may be made by more than one method and will be lengths or a combination of lengths and angles. The method chosen is a matter of convenience. If the motion of a particle is arbitrary its position at any given time can be described in terms of its distance, or **displacement**, from the origin of a rectangular coordinate system such as depicted in Figure 7.1.

The coordinate system in Figure 7.1 is said to be a **right-handed rectangular** system. Its axes are mutually perpendicular and are labelled in a certain way. If the system were rotated about the z-axis so the x-axis followed in the path of the y-axis, the z-axis would advance in the direction from O to z if it were a right-hand threaded screw. The y-axis in the figure is directed into the plane of the paper and the x–y plane is viewed from below. It is important to cultivate the habit of labelling axes in a systematic manner because it can simplify mathematical relationships in more advanced calculations.

The figure shows that the position of a moving particle, P, at any instant in time can be fully defined by the distances x_p, y_p and z_p measured along the appropriate axes to give the coordinate point (x_p, y_p, z_p). The point could equally well be defined using two lengths and an angle, or even one length and two angles. In certain circumstances it is convenient to take one or other of these approaches.

If the particle motion were constrained to lie on the surface of a cylinder, for example, it would be convenient to define its position in terms of a radius, r_p, a height, z_p and an angle θ_p as shown in Figure 7.2. Such coordinate systems, using two

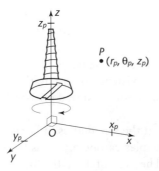

Figure 7.1 Rectangular coordinates.

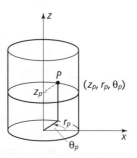

Figure 7.2 Cylindrical coordinates.

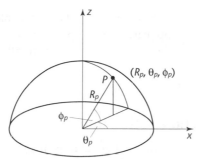

Figure 7.3 Spherical coordinates.

distances and an angle, are called **cylindrical** coordinate systems. Similarly, if the locus of the particle were constrained to lie on the surface of a sphere, perhaps a ship traversing the ocean, it would be more convenient to use one length, r_p, and two angles θ_p and ϕ_p (in this case corresponding to the radius of the earth, its latitude and its longitude). Such systems, not surprisingly, are called **spherical** coordinate systems (Figure 7.3).

If the particle were forced to follow a particular path, perhaps around the circumference at the top of a cylinder or along the equator of the globe, the motion would be **constrained**. And in both cases the motion would be **circular**. Furthermore, as it occurs wholly in the x–y plane, the motion may also be described as **planar**. Because planar motion is two-dimensional, only two coordinates are required to define a point, x_p and y_p if **Cartesian** coordinates are used, or r_p and θ_p if **polar** coordinates are used.

It is important to understand that any one of the rectangular, cylindrical or spherical coordinate systems can be used to define a point in three-dimensional space. Similarly, either Cartesian or polar coordinate systems can be used in two-dimensional space for planar motion. When the path of a particle is a straight line the situation simplifies still further and its location can be described by a single measurement of its displacement s_p from a reference point. This motion is called **rectilinear**.

We have discussed how the location of a particle can be defined at any instant in time. In kinematics we are interested in motions. They are generally described in terms of the changes of displacement in directions along the axes of the coordinate system in question. Motions are measured and described in terms of linear or angular displacements, velocities and accelerations. If the path follows a smooth curve, the particle experiences **curvilinear** motion and it is sometimes more convenient to measure quantities in directions tangential to, and at right angles to, its direction of travel at any given instant. These quantities are called **path variables**.

So far the motions of particles have been described with reference to their behaviour in space. They can also be described in relation to the manner in which they vary with time. For instance, some motions are repetitive, such as a bicycle pedal going round and round. If the motion is both regular and repetitive, and the period of time for the particle to complete each cycle of its motion is constant, it is known as **periodic** motion. Later chapters will describe motions in terms of their **temporal** behaviour.

In general, the more complicated the motion, the more difficult it is to describe exactly, and the less amenable it is to analysis. The principles of dynamics will therefore be introduced by their application to simple types of motion, starting with rectilinear motion and progressing to curvilinear motion. In all cases the motion will be planar.

7.2 Displacement, velocity and acceleration

Displacement, velocity and acceleration are part of most people's common experience. Every journey undertaken involves travelling a distance over a period of time. A measure of how much of the journey has been completed is the distance travelled since setting out – the displacement of the traveller from his or her point of departure. The time taken to complete the journey depends on the route taken and the speed of travel.

The idea of both speed and direction is embraced by the term velocity. The speed of a motor car may be 60 mph, but a complete specification of its velocity must include a reference to its direction, for example 60 mph travelling due north. Quantities that are specified in terms of magnitude and direction are called **vectors**; those specified by magnitude alone are called **scalars**. Speed is a scalar and velocity is a vector.

Journeys, for preference, tend to be undertaken at a more or less constant velocity since the shortest distance between two points is a straight line, and people tend to travel as fast as they comfortably can. This is more possible on a motorway than on country byroads. On a motorway people usually drive at a constant speed until they wish to carry out a manoeuvre, such as overtaking. In this case the driver changes the vehicle's direction of travel as he or she pulls out, generally increasing speed while overtaking, then changes direction again as he or she pulls in, perhaps reducing speed soon afterwards. In short, the driver accelerates to overtake. Acceleration involves changes in velocity; these could be changes in speed or direction, or both.

Most people not only have an intuitive understanding about velocity, displacement and acceleration, but also about the relationship between them. They will know that if they are travelling along the motorway without breaking the 70 mph speed limit, they will cover a maximum distance of 35 miles in a northerly direction in the next half-hour, providing they are travelling on the northbound carriageway. The most obvious relationship between speed, direction, distance travelled and time, is that **velocity is a measure of the distance covered in a given direction over a specified time interval.**

The actual distance covered depends on how much of the time was spent travelling at the speed limit. If the car had travelled with a constant speed of 70 mph throughout, it would have covered exactly 35 miles in the half-hour. If, as is more likely, the driver had needed to slow down for other traffic or roadworks, the average velocity over the half-hour would have been less than 70 mph and the distance covered would have been less than 35 miles. To put it another way, it would have taken longer than half an hour to cover the next 35 miles. Unless they are intending to break the speed limit, or travel by some other mode of transport, most people expect their average speed over their journey to be less than 70 mph, and reckon their journey time accordingly.

This approach is adequate for rough calculations of journey times but for many purposes, as will become apparent, it is necessary to have precise information not only about velocities but also about accelerations. Calculations of this type require a more rational and systematic understanding of the relationship between the various quantities. This will be explored further in the context of rectilinear motion.

7.3 Rectilinear motion

Rectilinear motion is motion in a straight line. To put it in a familiar context, consider a car travelling along a motorway. If we imagine the milometer reads s_1 at time t_1 and at some time later t_{end} it reads s_{end}, then the car will have travelled a distance $\Delta s = s_{end} - s_1$ in the period of time $\Delta t = t_{end} - t_1$. In Section 7.2 we chose a period of time Δt equal to half an hour and we showed that, if the average velocity $\hat{v}$ could be maintained at 70 mph, the distance covered in half an hour would be 35 miles.

$$\Delta s = \hat{v}\Delta t \qquad (7.1)$$

To calculate the distance travelled over a given time period, it is necessary to evaluate the average velocity. Its calculation requires a knowledge of the **time history** of the velocity. If the reading on the speedometer had been noted at regular intervals of time, a graph of the velocity plotted against the time would have given a **velocity–time curve**. The ease with which the average velocity can be determined depends on the nature of the velocity–time curve.

7.4 Constant velocity

The displacement, velocity and acceleration are all uniquely related to one another. The velocity–time (v–t) curve for a particle moving with constant velocity is shown in Figure 7.4, which also shows the corresponding curves for displacement and acceleration versus time. In this case the velocity v at any time t is equal to the average velocity $\hat{v}$; this is because v is constant with time. Figure 7.4 shows that the distance covered, equal to the area A_{v-t} under the velocity–time curve. To put this another way

$$\Delta s = \hat{v}\Delta t = 70 \times 0.5 = 35 \text{ miles}$$

or

$$\hat{v}\Delta t = A_{v-t} \rightarrow \hat{v} = A_{v-t}/\Delta t \qquad (7.2)$$

so the distance covered by a particle over a period of time is equal to the area under the velocity–time curve drawn for that period.

If we take Δt as a subinterval of the measured time, we can calculate intermediate distances. For example, the distance Δs_1 covered in the first 0.1 hour is given by the area under the v–t curve defined by

$$\Delta s_1 = \hat{v}_1\Delta t_1 = 70 \times 0.1 = 7 \text{ miles}$$

where $\hat{v}_1$ is the average velocity over the time interval Δt_1 taken as 0.1 hour. If the milometer showed an initial reading of 30 miles when the clock was started, the total displacement s_2 after 0.1 hour is

$$s_2 = s_1 + \Delta s_1 = 30 + 7 = 37 \text{ miles}$$

Because the velocity is constant, the distance covered for every interval of 0.1 hour will

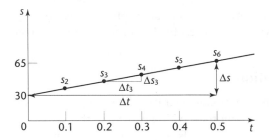

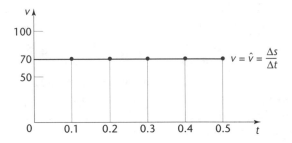

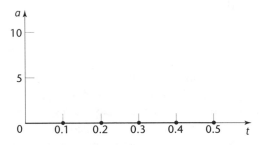

Figure 7.4 Displacement, velocity and acceleration plotted against time for a particle moving with constant velocity.

be 7 miles and the displacement after 0.2, 0.3, 0.4, and 0.5 hour will be 44, 51, 58 and 65 miles respectively. Over the time between $t_1 = 0$ and $t = 0.5$ hour, the displacement has changed from $s = 30$ to $s = 65$ miles, and the car has travelled another 35 miles.

The shape of the displacement–time (s–t) curve is fixed by the variation of the velocity with time. The slope of the s–t curve, m_{s-t}, in Figure 7.4 is the change in displacement, Δs, per change in time, Δt, or more formally, the rate of change of displacement with time, $\Delta s/\Delta t$. We have seen that for equal time intervals of 0.1 hour, the corresponding changes in displacement, Δs, are all equal to 7 miles. The slope of the s–t curve, m_{s-t}, over each interval is

$$\frac{\Delta s_1}{\Delta t_1} = \frac{\Delta s_2}{\Delta t_2} = \frac{\Delta s_3}{\Delta t_3} = \frac{\Delta s_4}{\Delta t_4} = \frac{\Delta s_5}{\Delta t_5} = \frac{7}{0.1} = 70 \text{ mph}$$

so the average slope of the s–t curve over any interval of time is equal to the average velocity over that interval.

This is restating Equation 7.1 as

$$\hat{v} = \frac{\Delta s}{\Delta t} \tag{7.3}$$

If the velocity is constant, the s–t curve is a straight line whose slope is equal to the magnitude of the velocity.

Acceleration is associated with changes in velocity. The velocity–time curve in Figure 7.4 shows there were no changes over the interval considered. The acceleration was therefore zero throughout.

Recap

■ The displacement s after a time Δt equals the initial displacement s_1 plus the change in displacement, Δs, occurring over the time Δt:

$$s = s_1 + \Delta s \tag{7.4}$$

■ The change in displacement equals the product of the average velocity $\hat{v}$ over the time Δt with Δt itself:

$$\Delta s = \hat{v}\Delta t \tag{7.5}$$

■ The average velocity is equal to the area A_{v-t} under the v–t curve for the period Δt, divided by Δt:

$$\hat{v} = A_{v-t}/\Delta t \tag{7.6}$$

■ Combining Equations 7.4 to 7.6, the displacement after a time Δt can be found in terms of the average velocity or in terms of the area under the v–t curve:

$$s = s_1 + \hat{v}\Delta t \tag{7.7}$$

$$s = s_1 + A_{v-t} \tag{7.8}$$

■ Finally, the average velocity over the time interval Δt can be found in terms of the change in displacement over the interval Δs divided by Δt, or from the average slope m_{s-t} of the s–t curve over the interval Δt:

$$\hat{v} = \frac{\Delta s}{\Delta t} \tag{7.9}$$

$$\hat{v} = \hat{m}_{s-t} \tag{7.10}$$

The results presented so far have been argued from the case of constant velocity. They also hold for steadily varying velocity, and for the general case of arbitrarily varying velocity.

7.5   Constant acceleration

We will now consider the simplest type of rectilinear motion for which the velocity varies, namely motion with constant acceleration. The s–t, v–t and a–t curves for a particle moving with constant acceleration are shown in Figure 7.5. These curves could describe the motion of a vehicle travelling on a motorway, but in this example the driver accelerates steadily from 50 to 70 mph.

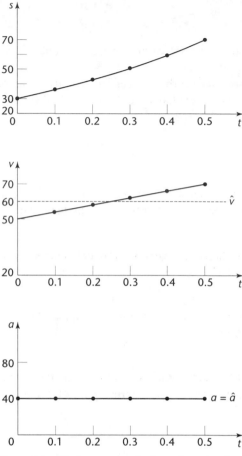

Figure 7.5 Displacement, velocity and acceleration plotted against time for a particle moving with constant acceleration.

Notice how the curves in Figure 7.5 are quite different from the curves in Figure 7.4. The s–t curve is now a gentle curve with increasing slope instead of a straight line, although the scale of the ordinate axis has to be magnified to see this clearly. The velocity increases linearly with time, instead of remaining constant, and the acceleration is now finite (non-zero). And notice that the average velocity over the half-hour is less than 70 mph; the displacement changes from 30 to 60 miles, giving a total change of displacement of 30 miles rather than the 35 miles obtained in Section 7.4.

For constant velocity we saw that important relationships between displacement and velocity could be deduced from the slopes and areas of the s–t and v–t curves. By examining the v–t curve in this example we can see that the average velocity over the half-hour, $\hat{v} = \frac{1}{2}(50 + 70) = 60$ mph. And the area under the curve, $A_{v-t} = \frac{1}{2}(50 + 70) \times 0.5 = 30$ miles, consistent with Equation 7.6. Let us now check the validity of Equations 7.7 and 7.8.

$$s = s_1 + \hat{v}\Delta t = 30 + 60 \times 0.5 = 60 \text{ miles}$$

$$s = s_1 + A_{v-t} = 30 + 30 = 60 \text{ miles}$$

The equations are self-consistent, and the results agree with the s–t curve.

Inspection of the v–t curve shows that over the half-hour time interval Δt the change in velocity Δv is $70 - 50 = 20$ mph. The area under the a–t curve for the same time interval is (40×0.5) mph. We already know the change in displacement, Δs, is given by the area A_{s-t} under the s–t curve. **In the same way the change in velocity, Δv, is equal to the area A_{a-t} under the a–t curve for the time interval Δt.** Expressing this algebraically, we have

$$\Delta v = A_{a-t} = \hat{a}\Delta t \tag{7.11}$$

where the acceleration a is equal to the average acceleration $\hat{a}$; this is because a remains constant over the time interval.

And just as Equation 7.1 could be restated to give the average velocity in terms of the change in displacement per time interval (Equation 7.9) or in terms of the slope of the s–t curve (Equation 7.10), so Equation 7.11 can be restated in the following forms:

$$\hat{a} = \frac{\Delta v}{\Delta t} \tag{7.12}$$

$$\hat{a} = \hat{m}_{v-t} \tag{7.13}$$

where $\hat{m}_{v-t}$ is the average slope of the v–t curve. **The average acceleration over any time interval Δt is equal to the average rate of change of velocity with time, $\Delta v/\Delta t$, which is also equal to the average slope of the v–t curve, $\hat{m}_{v-t}$, over that interval Δt.** In Figure 7.5 the change in velocity, Δv, in half an hour is $70 - 50 = 20$ mph, so $\Delta v/\Delta t$ is $20/0.5 = 40$ mph^2, which is also the slope of the v–t curve and the value of $\hat{a}$ shown on the s–t curve.

Since the acceleration is constant, the slope of the v–t curve is constant. The velocity therefore increases steadily by 4 mph for each of the five 0.1 hour intervals, giving a total increase of 20 mph over the half-hour. The slope $\Delta v/\Delta t$ has the same value whether Δt is 0.5 hour (when $\Delta v/\Delta t = 20/0.5$) or 0.1 hour (when $\Delta v/\Delta t = 4/0.1$).

The same is not true for the slope of the s–t curve. The change in displacement, Δs, over half an hour is $60 - 30 = 30$ miles. The average velocity $\hat{v}$, as seen on the v–t curve, is $\frac{1}{2}(50 + 70) = 60$ mph. The average slope of the s–t curve, $\Delta s/\Delta t = 30/0.5$ and does indeed equal the average velocity $\hat{v}$ over the half-hour. However, the change in displacement, Δs_1, over the first 0.1 hour is 5.2 miles, so $\Delta s_1/\Delta t = 52$ mph, which is the average velocity over that period. We know this is true because the initial velocity is 50 mph and it increases by 4 mph for every 0.1 hour.

In the last 0.1 hour of the half-hour, the displacement changes by 6.8 miles, giving an average velocity over that period of 68 mph. The changing slope of the s–t curve reflects the way the velocity changes over the half-hour. At the beginning of the motion the velocity is low and the distance covered per unit time is small, so the slope of the s–t curve is relatively small. At the high end, the reverse is true and the slope is relatively large. Average velocities over subintervals of the motion can be higher or lower than the average velocity over the full half-hour.

➤ ## 7.6 Equations for constant acceleration

So far we have deduced some general equations in terms of average velocities and accelerations. In the **special case of constant acceleration**, because the velocity v varies as a straight line with time, the average velocity $\hat{v}$ over any given time interval can be calculated exactly from its initial velocity v_1, and its final velocity v_2, over that period:

$$\hat{v} = \frac{1}{2}(v_2 + v_1) \tag{7.14}$$

In Figure 7.5, as we have seen, the average velocity over the first 0.1 hour is exactly $\frac{1}{2}(50 + 54) = 52$ mph. We can use this relationship, together with the fact that $\hat{a} = a$ for all time, to derive equations in terms of instantaneous velocities (velocities at particular instants in time) rather than average velocities.

Let us consider a case where the displacement $s_1 = 0$ when the time $t_1 = 0$, then $\Delta s = (s_2 - s_1) = s_2$ and $\Delta t = (t_2 - t_1) = t_2$. Equation 7.5 can now be written as

$$s_2 = \hat{v}t_2$$

giving

$$s_2 = \frac{1}{2}(v_2 + v_1)t_2 \tag{7.15}$$

Similarly, Equation 7.11 can be written as

$$v_2 - v_1 = at_2$$

giving

$$v_2 = v_1 + at_2 \tag{7.16}$$

Turning now to Equation 7.15, it can be seen that

$$\frac{1}{2}(v_2 + v_1) = \frac{s_2}{t_2}$$

and bearing in mind that

$$v_2 - v_1 = at_2 \tag{7.17}$$

Multiplying the right-hand side of one equation by the other, and equating to the product of the left-hand sides, leads to

$$\frac{1}{2}(v_2 + v_1)(v_2 - v_1) = \left(\frac{s_2}{t_2}\right)(at_2)$$

$$v_2^2 - v_1^2 = 2as_2$$

$$v_2^2 = v_1^2 + 2as_2 \tag{7.18}$$

Finally, if we multiply Equation 7.17 by $\frac{1}{2}t_2$ to give

$$\frac{1}{2}at_2^2 = \frac{1}{2}(v_2 - v_1)t_2$$

and subtract it from Equation 7.15, we get

$$s - \frac{1}{2}at_2^2 = v_1 t_2$$

giving

$$s = v_1 t_2 + \frac{1}{2}at_2^2 \qquad (7.19)$$

Recap

▪ If a particle is moving in a straight line with constant acceleration, the temporal relationship between its displacement, velocity and acceleration can be expressed by the following equations where the suffixes denote initial and final conditions, and the initial displacement $s_1 = 0$ when the time $t_1 = 0$.

$$s_2 = \frac{1}{2}(v_1 + v_2)t_2$$

$$s_2 = v_1 t_2 + \frac{1}{2}at_2^2$$

$$v_2 = v_1 + at_2$$

$$v_2^2 = v_1^2 + 2as_2$$

▪ The equations have been derived for the special case of rectilinear motion with constant acceleration and $t_1 = s_1 = 0$, and are only valid under these conditions.

The equations connect the five quantities v_1, s_2, v_2, t_2 and a. Each equation is composed of a different set of four out of the five. A systematic way of tackling problems using these equations is to make a list of the information then to choose the equation that connects the known quantities with the desired unknown.

Example 7.1

Suppose that a motorist driving along a straight motorway with constant acceleration looks at the speedometer and notes it is reading 80 kph. Half an hour later the motorist is travelling at 113 kph, according to the speedometer. How far has the motorist travelled in the meantime, and what has been his or her acceleration?

Solution   There are seven quantities of interest in this problem, the initial time, displacement and velocity (t_1, s_1, v_1), their corresponding final values (t_2, s_2, v_2) and the acceleration a. It is convenient to take the initial values $t_1 = s_1 = 0$ and to use the equations involving the remaining quantities, as derived earlier. Each of these equations contains four of the quantities v_1, t_2, s_2, v_2 and a. If we know three of them, as we do here, it is possible to find two separate equations to determine the remaining two unknowns.

A systematic approach to these problems is to write down the initial and final conditions, noting which quantities are specified by the initial data.

Initial conditions

$t_1 = 0$, $s_1 = 0$, $v_1 = 80$ kph

Final conditions

$t_2 = 0.5$ hour, $s_2 = ?$ $v_2 = 113$ kph

Acceleration

$a = ?$

In order to answer the first part of the question, we need an equation connecting the three quantities we know (v_1, v_2, t_2) with the quantity we wish to find (s_2). The appropriate equation is obviously Equation 7.15:

$$s_2 = \frac{1}{2}(v_2 + v_1)t_2$$
$$= \frac{1}{2}(113 + 80) \times 0.5$$
$$= \textbf{48.25 km}$$

Looking now for an equation that connects v_1, v_2, t_2 with the acceleration, we see the appropriate choice is Equation 7.16:

$$v_2 = v_1 + at_2$$
$$a = \frac{v_2 - v_1}{t_2}$$
$$= \frac{113 - 80}{0.5}$$
$$= \textbf{66 kph}^2$$

In fact it would have been possible to obtain an answer for the acceleration using any of the equations in which it appears, because we have calculated a value for s_2, which gives us an additional piece of data. Using Equation 7.18, for example, we have

$$v_2^2 = v_1^2 + 2as_2$$
$$a = \frac{v_2^2 - v_1^2}{2s_1}$$
$$= \frac{113^2 - 80^2}{2 \times 48.25}$$
$$= \textbf{66 kph}^2$$

However, obtaining a correct answer by this approach relies on calculating the correct value for s_2. A wrong answer for s_2 would lead to an incorrect value for a. It is always desirable to work from the initial data whenever possible.

➤ **Example 7.2**

A particle moves with rectilinear motion and has an initial velocity of $80 \, \text{m s}^{-1}$. For the first 5 s it has zero acceleration, after which it is slowed down by a resistance force giving it a constant acceleration of $-15 \, \text{m s}^{-2}$. Calculate how far it has travelled after 10 s and its velocity at that time.

Solution The first thing to notice is that the acceleration varies over the 10 s of motion, so the equations of motion with constant acceleration cannot be applied simply to the conditions at the beginning and end of the motion. However, the acceleration is constant for the first 5 s, when it is zero, and also for the second 5 s. It is convenient therefore to divide the problem into two parts.

First 5 s

During the first 5 s there is no acceleration, so the velocity does not change.
Initial conditions

$$t_1 = 0 \quad s_1 = 0 \quad v_1 = 80 \text{ m s}^{1} \quad a = 0$$

Final conditions

$$t_2 = 5 \text{ s} \quad s_2 = ? \quad v_2 = 80 \text{ m s}^{-1} \quad a = 0$$

The only unknown is the displacement, s_2. In Equation 7.15 the displacement is the subject of an equation connecting it with velocity and time. The displacement over the first 5 s can therefore be found directly.

$$s_2 = \frac{1}{2}(v_2 + v_1)t_2 = \frac{1}{2}(80 + 80) \times 5 = 400 \text{ m} \tag{7.20}$$

Second 5 s

During the second 5 s the particle is slowed down by a retarding force with a constant deceleration of 15 m s^{-2}. Resetting the time datum to zero to allow the use of the equations of motion for constant acceleration (where a takes a negative sign for deceleration) leads to the following initial and final conditions:
Initial conditions

$$t_1 = 0 \quad s_1 = 0 \quad v_1 = 80 \text{ m s}^{-1} \quad a = -15 \text{ m s}^{-2}$$

Final conditions

$$t_2 = 5 \text{ s} \quad s_2 = ? \quad v_2 = ? \quad a = -15 \text{ m s}^{-2}$$

Examination of Equations 7.15 to 7.19 shows that the only one we can solve directly is Equation 7.16:

$$v_2 = v_1 + at_2 = 80 + (-15) \times 5 = 5 \text{ m s}^{1} \tag{7.21}$$

Now that v_2 is known, s_2 can be calculated from any of the equations in which it appears; the easiest is Equation 7.15:

$$s_2 = \frac{1}{2}(v_1 + v_2)t_2 = \frac{1}{2}(80 + 5) \times 5 = 212.5 \text{ m}$$

Over the whole period of 10 s the particle has travelled a total distance of 612.5 m and its final velocity is 5 m s^{-1}.

➤ 7.7 Problems

1. In Figure 7.6 the y-axes all point into the plane of the page. Which sets of coordinates are right-handed?

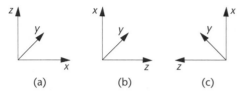

(a)   (b)   (c)

Figure 7.6

2. (a) If a point (x, y, z) is given by $(1, 1, \sqrt{2})$ what are the corresponding points (r, θ, z) and (R, θ, ϕ) in cylindrical and spherical coordinate systems with the same origin?

   (b) Similarly, if a point (r, θ, z) is given by $(2, \pi/6, 2)$, what are the points (x, y, z) and (R, θ, ϕ)?

   (c) Finally, if a point (R, θ, ϕ) is given by $(4\sqrt{3}, \pi/3, \pi/6)$, what are the points (x, y, z) and (r, θ, z)?

3. A person sets off on a car journey at 09:00 hrs and averages a speed of 45 mph until she turns on to a motorway at 09:50 hrs when her milometer reads 10 000 miles. She reaches her turnoff at 12:20 hrs, at which point her milometer reads 10 162.5 miles. She completes the final 23 mile stretch of her journey at an average speed of 46 mph.

   (a) What does her milometer read at the beginning of her journey?

   (b) What is her average speed on the motorway?

   (c) At what time does she reach her destination?

   (d) What is her average speed over her journey?

4. In an experiment, the kinematics of a particle were being recorded. It was noted that after travelling 10 m it then moved in a straight line travelling with a constant velocity of v m s^{-1} for T s before being brought to an abrupt halt. During this latter period the magnitude of the area under the v–t curve was 65 and that of the slope of the s–t curve was 5, both measured in standard SI units. Calculate v, T and the distance the particle travelled.

5. A train is travelling at 90 km h^{-1}. Its brakes are applied so that it comes to a halt in a station with a steady deceleration, a, over a distance of 200 m. What is the value of a, and how long does the train take to stop?

6. A jet aircraft taxis down the runway at a near constant acceleration of 6 m s^{-2}. Calculate how far it taxis, and the time taken, before take-off if the lift-off speed is 216 km h^{-1}.

7. A small rocket is launched vertically and travels with a constant acceleration of 12 m s^{-2} until, after 6 s, its fuel is exhausted. Assuming that further progress is retarded only by gravity, calculate the maximum height it achieves. (Hint: with no fuel the rocket speed decreases to zero, when it reaches its maximum height.)

8. Design specifications for a high speed lift in a high rise building limit its acceleration/deceleration to ± 3.5 m s^{-2} and its speed to 7 m s^{-2}. Calculate the minimum time taken for the lift to come to rest at the 17th floor having started at the bottom floor, a total distance of 84 m, and also the distance it travels at constant speed.

8 Curves, slopes and areas

The slopes and areas of the displacement, velocity and acceleration time curves are fundamental to establishing the relationship between displacement, velocity and acceleration. It is therefore worthwhile spending some time considering these parameters and ways of quantifying them.

8.1 Curves

We will consider first the curves shown in Figure 8.1. They are another family of

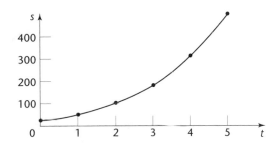

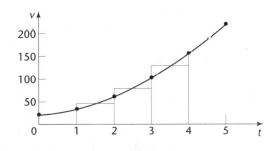

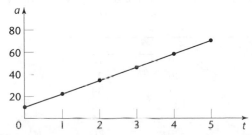

Figure 8.1 Displacement, velocity and acceleration plotted against time for a projectile.

graphs showing the variation of displacement, velocity and acceleration with time and perhaps describing the initial kinematics of a projectile. In this example, unlike the previous ones, the acceleration changes with time, so the displacement–time and the velocity–time curves are truly curves in the non-mathematical sense of the word.

The a–t curve is a straight line connecting the points whose coordinates $(a, t) = (10, 0), (22, 1), (34, 2), (46, 3), (58, 4)$ and $(70, 5)$. The acceleration increases by $12\,\mathrm{m\,s^{-2}}$ every second. The slope of the graph, $\Delta a / \Delta t$, at any point on the curve can be found from any pair of points and will be constant. Taking the first pair of points:

$$\frac{\Delta a}{\Delta t} = \frac{22 - 10}{1 - 0} = 12 \ \mathrm{m\,s^{-3}}$$

and taking the fifth and sixth points:

$$\frac{\Delta a}{\Delta t} = \frac{70 - 58}{5 - 4} = 12 \ \mathrm{m\,s^{-3}}$$

Straight lines can always be represented mathematically by an equation of the form

$$y = mx + c_0 \tag{8.1}$$

where the value of y changes with the value of x, m is the slope of the line and c_0 is the value of y when $x = 0$. So the a–t curve of Figure 8.1 can be represented by the equation

$$a = 12t + 10 \quad (\mathrm{m\,s^{-2}}) \tag{8.2}$$

The coordinates of any of the points lying on the a–t curve satisfy Equation 8.2: $a = 10\,\mathrm{m\,s^{-1}}$ when $t = 0$, which corresponds to the coordinates of the first point $(10, 0)$. Similarly, when $t = 5\,\mathrm{s}$, $a = 12 \times 5 + 10 = 70\,\mathrm{m\,s^{-2}}$, which corresponds to the coordinates of the sixth point, $(70, 5)$.

Variation of the acceleration with time can be represented by a discrete relationship either using numbers (a, t), or graphically using points, showing the acceleration at exact moments in time. Alternatively it can be represented by a continuous relationship using Equation 8.2 or graphically by a continuous curve, showing the value of the acceleration at any time t for which the equation or curve remains valid. Equation 8.2 can be seen as a mathematical representation of the a–t curve shown in Figure 8.1, or vice versa, the a–t curve may be seen as a graphical representation of Equation 8.2. Both describe exactly the variation of the acceleration with time.

Similarly, the variation of the displacement and the velocity with time can also be described by numbers, as shown in Table 8.1, or by the following equations:

Table 8.1

t (s)	s (m)	v (m s^{-1})	a (m s^{-2})
0	25	20	10
1	52	36	22
2	101	64	34
3	184	104	46
4	313	156	58
5	500	220	70

$$s = 2t^3 + 5t^2 + 20t + 25 \tag{8.3}$$

$$v = 6t^2 + 10t + 20 \tag{8.4}$$

The relationships between displacement and time and between velocity and time can be represented graphically by discrete points or as continuous curves. The numbers shown in Table 8.1 satisfy Equations 8.3 and 8.4.

8.2 Slopes and differentiation

The slope of the a–t curve could be found by taking pairs of points and calculating the change in acceleration over a given time interval. This gave a correct answer no matter which pair was taken because the rate of increase of a with time does not change with time – the slope is constant. But the displacement increases ever more rapidly as time goes by. The rate of change of displacement with time, or the slope of the s–t curve, itself changes with time. In this case, calculating slopes from pairs of points gives approximate values of the slope midway between the points.

By taking the first and third points for the displacement shown in Table 8.1, the slope of the s–t curve when $t = 1$ s can be estimated as

$$\frac{\Delta s}{\Delta t} = \frac{101 - 25}{2 - 0} = 38 \text{ m s}^{-1}$$

Similarly the slopes at times $t = 2$, 3 and 4 s can be estimated using pairs of points on either side of the time being considered.

Section 7.4 showed how the slope of the s–t curve approximated over an interval gives the average velocity over that interval (see Equation 7.9). The average velocity over the interval $t = 0$ s to $t = 2$ s is therefore 38 m s^{-1}. Table 8.1 shows that the instantaneous velocity at the midpoint of the interval, when $t = 1$ s, is 36 m s^{-1}. The average velocity over the interval is therefore a good approximation to the instantaneous velocity at the midpoint of that interval. How good an approximation depends on the chosen magnitude of Δt. For example, if the slope were approximated from displacements at 0.5 s and 1.5 s, which Equation 8.3 gives as 36.5 m and 73.0 m, then

$$\frac{\Delta s}{\Delta t} = \frac{73.0 - 36.5}{1.5 - 0.5} = 36.5 \text{ m s}^{-1}$$

The smaller the interval, the more closely $\Delta s/\Delta t$ approximates the actual slope of the s–t curve, and the nearer the average velocity approaches the instantaneous velocity. In fact, the instantaneous velocity equals the average velocity as the interval Δt becomes infinitesimally small, i.e. as Δt approaches zero.

Several points emerge from this discussion:

➤ The slope of the s–t curve is a measure of how fast the displacement s changes with time t. In other words, it is the rate of change of s with respect to t.
➤ The slope can be approximated over an interval of time by $\Delta s/\Delta t$, which gives the average velocity $\hat{v}$ over that interval.
➤ The slope at an instant in time equals the approximate slope over an interval of time as that interval becomes infinitesimally small. It may be written

$$\frac{d}{dt}(s) = \lim_{\Delta t \to 0} \frac{\Delta s}{\Delta t} = \frac{ds}{dt}$$

where d/dt may be thought of as a shorthand symbol meaning the rate of change of some quantity with respect to time.

➤ The instantaneous velocity at any time is the rate of change of displacement with respect to time, which in turn is the slope of the s–t curve at that time, i.e.

$$v = \frac{ds}{dt}$$

The term ds/dt is often known as the differential of the displacement s with respect to time t. Looking back to Equation 7.12, it will come as no surprise that acceleration can also be expressed in terms of a differential:

$$a = \lim_{\Delta t \to 0} \frac{\Delta v}{\Delta t} = \frac{d}{dt}(v) = \frac{dv}{dt}$$

If some quantity of interest, y say, can be expressed in terms of another quantity, x say, the differential of y with respect to x, dy/dx, can often be obtained by the application of a few relatively simple rules.

Suppose

$$y = cx^n$$

then

$$\frac{dy}{dx} = ncx^{n-1} \tag{8.5}$$

where c is a constant (a fixed number).

We can apply this rule to the case of constant velocity in Section 7.4. We can see how the variation of displacement with time is given by a straight line of the same form as Equation 8.1. The value of s when $t = 0$ is 30 miles, so $c_0 = 30$, and it increases to 65 miles in half an hour, so $m = (65 - 30)/0.5 = 70\,\text{mph}^2$. The equation for the s–t curve of Figure 7.4 is therefore given by

$$s = 70t + 30 \quad \text{(miles)} \tag{8.6}$$

If we want to find the differential of s with respect to t, we require to know the rate at which the terms on the right-hand side of the equation change with time. The second term is easy. It is a constant that doesn't change with time, so its rate of change with time is zero. To deal with the first term we can use the rule given by Equation 8.5 with $y = s$, $x = t$, $c = 70$ and $n = 1$, then

$$s = 70t^1 + 30$$

$$\frac{ds}{dt} = 1 \times 70 \times t^0 + 0 = 1 \times 70 \times 1 + 0 = 70 \quad \text{(mph)}$$

Furthermore, since $a = dv/dt$, we can also find the acceleration at any given time by differentiating Equation 8.3 with respect to time:

$$v = 6t^2 + 10t^1 + 20t^0$$

$$a = \frac{dv}{dt} = 2 \times 6 \times t^1 + 1 \times 10 \times t^0 + 0 \times 20 \times t^{-1}$$

$$a = 12t + 10 \quad (\mathrm{m\,s^{-2}})$$

This is identical to Equation 8.2, which we know to be the equation of a straight line. The acceleration at time $t = 5\,\mathrm{s}$, $a = 12 \times 5 + 10 = 70\,\mathrm{m\,s^{-2}}$.

So the a–t equation can be derived by differentiating the v–t equation, which is the same as differentiating the s–t equation twice, all differentiation with respect to time. Various forms of notation are used to indicate differentiation with respect to time. Two common forms are

$$v = \frac{ds}{dt} \quad v = \dot{s} \tag{8.7}$$

$$a = \frac{dv}{dt} \quad a = \dot{v}$$

$$= \frac{d}{dt}\left(\frac{ds}{dt}\right) \tag{8.8}$$

$$= \frac{d^2 s}{dt^2} \quad a = \ddot{s}$$

Differentiation is a very powerful tool that is used in all branches of mathematics, science and engineering. Its value in dynamics increases as the problems to be tackled become more intractable. Although simple problems can often be solved without recourse to differentiation, an appreciation of it at an early stage will reap rewards later on.

8.3 Areas and integration

The area under the v–t curve of Figure 8.1 between the limits $t = t_0$ and $t = t_N$ can be estimated by splitting it up into N rectangular strips of width $\Delta t = (t^N - t_0)/N$ and of height v_i, where v_i is the velocity at the midpoint of the ith strip and $i = 1, \ldots, N$. For example, the figure has $N = 3$, $t_0 = 1\,\mathrm{s}$ and $t_N = 4\,\mathrm{s}$. The strip width, $\Delta t = (4 - 1)/3 = 1\,\mathrm{s}$, $v_1 = 48.5\,\mathrm{m\,s^{-1}}$, $v_2 = 82.5\,\mathrm{m\,s^{-1}}$ and $v_3 = 128.5\,\mathrm{m\,s^{-1}}$. The approximate area under the curve A_{v-t}^* is given by

$$A_{v-t}^* = v_1 \Delta t + v_2 \Delta t + v_3 \Delta t$$

$$= 48.5 \times 1 + 82.5 \times 1 + 128.5 \times 1 \tag{8.9}$$

$$= 259.5\,\mathrm{m}$$

where the asterisk signifies that A_{v-t}^* is an estimate of the actual area A_{v-t}.

Figure 8.1 shows that the strips include some area above the curve and miss out some area below the curve. To some extent these two errors in estimating the true area under the curve may be expected to cancel. The smaller the width of the strip, the smaller the error, but the larger the number of strips required to cover the same area. For large numbers of strips it is more convenient to write Equation 8.9 as

$$A^*_{v-t} = v_1 \times \Delta t + v_2 \times \Delta t + \ldots + v_N \Delta t$$

$$= \sum_{i=1}^{N} v_i \Delta t \quad i = 1, \ldots, N \tag{8.10}$$

As the interval of three seconds is divided into a larger and larger number of strips, the approximate value of the area under the curve, A^*_{v-t}, changes more and more slowly as it converges towards the true area, A_{v-t}. In this example, if six strips ($N = 6$) of width $\Delta t = 0.5$ s are used, the approximate area, A^*_{v-t}, is 260.625 m. If twelve strips of width $\Delta t = 0.25$ s are used, A^*_{v-t} is 260.906 25 m. The true value of the area under the curve, A_{v-t}, as we shall see, is 261 m. So already, using only twelve strips, the estimated area is within 0.036% of the true area.

The true area under the curve of velocity plotted against time over the interval from t_0 to t_N is the limiting value obtained by summing the elemental areas as their width Δt becomes infinitesimally small. This can be written as

$$A_{v-t} = \lim_{\Delta t \to 0} \sum_{i=1}^{N} v_i \Delta t \quad i = 1, 2, \ldots, N$$

where

$$\Delta t = \frac{t_N - t_0}{N} \tag{8.11}$$

As noted in Section 7.5, we are also interested in evaluating the area under the acceleration–time curve, and the same approach can be used. In fact, this approach can be used to evaluate the area under the curve of any quantity y that varies with another quantity x over some given interval of x, and Equation 8.11 can be generalized to

$$A_{y-x} = \lim_{\Delta x \to 0} \sum_{i=1}^{N} y_i \Delta x \quad i = 1, \ldots, N \tag{8.12}$$

Like the slopes of curves, the approximate areas under curves can be found graphically, or can be estimated from a series of discrete points. And just as for slopes, if the equations of the curves are known, a mathematical process can be used to find the exact areas. If we have an equation for the curve of y plotted against x, the value of y varies with the value of x in a precise way that is dictated by the equation. The variable y is called a function of x, written mathematically as

$$y = f(x)$$

The value of y at the point when $x = x_i$ is written as

$$y_i = f(x_i)$$

If we introduce this notation into Equation 8.12

$$A_{y-x} = \lim_{\Delta x \to 0} \sum_{i=1}^{N} f(x_i) \Delta x \quad i = 1, \ldots, N \tag{8.13}$$

The summation given by the right-hand side of Equation 8.13 is known as the definite integral of the function $f(x)$ between the limits $x = a$ and $x = b$ and is represented by the notation

$$\int_a^b f(x) \, dx = \lim_{\Delta x \to 0} \sum_{i=1}^{N} f(x_i)\Delta x \quad i = 1, \ldots, N \tag{8.14}$$

In this case $a = x_0$ and $b = x_N$.

The definite integral of $f(x)$ is equal to the area under the curve between the limits $x = a$ and $x = b$. The symbol $\int_a^b f(x) \, dx$ is simply a more precise way of writing A_{y-x}. If the equation for $f(x)$ is known, the definite integral can often be calculated by integration using a few simple rules.

When we were looking at differentiation, we considered this function:

$$y = f(x) = cx^n$$

The first step in obtaining the definite integral of $f(x)$ is to raise the power of x by 1 and to divide the function by $n + 1$:

$$\int_a^b f(x) \, dx = \int_a^b cx^n \, dx$$

$$= \left[\frac{cx^{n+1}}{n+1}\right]_a^b \tag{8.15}$$

The square brackets indicate that the function has been integrated but the limits a and b have to be substituted into the new function before obtaining the final answer. The next step, then, is to subtract the new function evaluated with $x = a$ from the new function evaluated with $x = b$:

$$\left[\frac{cx^{n+1}}{n+1}\right]_a^b = \frac{cb^{n+1}}{n+1} - \frac{ca^{n+1}}{n+1}$$

The outcome of the process of obtaining the definite integral is a single number equal to the area under the curve of $f(x)$ between the limits $x = a$ and $x = b$.

To illustrate the process we will consider the function $f(x) = 12x$ integrated between the limits $x = 0$ and $x = 5$.

$$\int_a^b f(x) \, dx = \int_0^5 12x \, dx = \int_0^5 12x^1 \, dx$$

$$= \left[\frac{1 \times 12x^{(1+1)}}{(1+1)}\right]_0^5 = \left[\frac{12x^2}{2}\right]_0^5$$

$$= \frac{12 \times 5^2}{2} - \frac{12 \times 0^2}{2} = 150 \tag{8.16}$$

Examining Equation 8.2, we can see that we have integrated the first term of the acceleration–time curve in Figure 8.1. Notice how the a–t curve is a straight line, so the area under it, A_{a-t}, between the times $t = 0$ and $t = 5$ s can be easily calculated. Since

$$a = 12t + 10 \quad (\text{m s}^{-2})$$

when $t = 0$, $a = 10\,\mathrm{m\,s^{-2}}$ and when $t = 5\,\mathrm{s}$, $a = 70\,\mathrm{m\,s^{-2}}$. The area under the a–t curve can be split into a triangle of height $60\,\mathrm{m\,s^{-2}}$ and base $5\,\mathrm{s}$, and a rectangle of height $10\,\mathrm{m\,s^{-2}}$ and base $5\,\mathrm{s}$. The total area is therefore

$$A_{a-t} = \frac{1}{2} \times 60 \times 5 + 10 \times 5 = 200\ \mathrm{m\,s^{-1}}$$

If we now integrate the complete a–t equation between the limits $t = 0$ and $t = 5\,\mathrm{s}$, we can expect to obtain the answer of $200\,\mathrm{m\,s^{-1}}$, since the integral is simply the area under the a–t curve between those limits.

Repeating the integration procedure for the variable t instead of x, we get

$$a = f(t) = 12t + 10 = 12t^1 + 10t^0$$

Therefore

$$\int_0^5 f(t)\,\mathrm{d}t = \int_0^5 (12t^1 + 10t^0)\,\mathrm{d}t$$

$$= \left[\frac{12t^{(1+1)}}{(1+1)} + \frac{10t^{(0+1)}}{0+1} \right]_0^5$$

$$= \left[\frac{12t^2}{2} + \frac{10t}{1} \right]_0^5$$

$$= \left(\frac{12 \times 5^2}{2} - \frac{12 \times 0^2}{2} \right) + \left(\frac{10 \times 5}{1} - \frac{10 \times 0}{1} \right)$$

$$= 200\ \mathrm{m\,s^{-1}} \tag{8.17}$$

In the case we have been considering

$$\int_0^5 a\,\mathrm{d}t = \int_0^5 f(t)\,\mathrm{d}t = A_{a-t} = 200\ \mathrm{m\,s^{-1}}$$

which is what we expected. Notice in passing that the first term in brackets equals the area of the triangle, and the second term the area of the rectangle under the a–t curve.

To demonstrate that integration can also be used to find areas under curves that are not straight lines, we will now apply it to the v–t curve in Figure 8.1 and described by Equation 8.4. When we divided its area into twelve thin strips of $0.25\,\mathrm{s}$ width, we estimated that the area A_{v-t}^* under the curve between the limits $t = 1\,\mathrm{s}$ and $t = 4\,\mathrm{s}$ equalled $260.906\,25\,\mathrm{m}$. The equation of the v–t curve is

$$v = f(t) = 6t^2 + 10t + 20 \quad (\mathrm{m\,s^{-1}})$$

Therefore

$$\int_1^4 v \, dt = \int_1^4 (6t^2 + 10t + 20) \, dt$$

$$= \left[\frac{6t^3}{3} + \frac{10t^2}{2} + \frac{20t}{1} \right]_1^4$$

$$= \left(\frac{6 \times 64}{3} - \frac{6 \times 1}{3} \right) + \left(\frac{10 \times 16}{2} - \frac{10 \times 1}{2} \right) + \left(\frac{20 \times 4}{1} - \frac{20 \times 1}{1} \right)$$

$$= 261 \text{ m} \tag{8.18}$$

The area A^*_{v-t} estimated using twelve strips is very nearly equal to the value of the definite integral, which is in fact the exact area.

▶ 8.4   Using integration in kinematics

The discussion on integration has concentrated so far on its use for calculating areas. We shall return to this shortly, but first we consider another aspect of the subject. If we examine the terms on the right-hand side of Equation 8.18, we notice that after integration but before applying the limits (terms in brackets on the following line) they become equal to the first three terms of Equation 8.3, relating distance and time. The only difference between the right-hand side of Equation 8.3 and the terms in the brackets is a constant, in this case equal to 25. Therefore, one can write

$$s = \int v \, dt = 2t^3 + 5t^2 + 20t + c \tag{8.19}$$

where c is termed the constant of integration. The term on the left-hand side of the equation is known as the indefinite integral; this is because the process of integration has been carried out but no limits have been applied, so the value of c (and therefore of the integral itself) is unknown.

It follows that, if the equation of the v–t curve is known, the equation for the s–t curve can be found by integrating it, providing the displacement s_0 at a given time t_0 is also known. Looking back to Table 8.1, we can take any pair of points (s_0, t_0), say $(52, 1)$, to calculate the value of c in Equation 8.19.

$$52 = 2 \times 1^3 + 5 \times 1^2 + 20 \times 1 + c$$

$$c = 52 - 27 = 25$$

Although the result of taking the definite integral is a single number representing an area, the result of taking the indefinite integral is an equation representing a family of curves characterized by the value of the constant c. Further information, the point (s_0, t_0), is required to define the required member of the family. In Section 8.2 we showed that, if we knew the displacement s as a function of t, we could find the equations for both the velocity and the acceleration by successive differentiation. In the case of integration, if we know the equation for the acceleration, we can integrate it once to find the equation for the velocity, and again to find the equation for the displacement. In each case, however, some extra information is required to determine the constant of integration.

Having discussed differentiation and integration, and how they relate to each other, we will now consider how they can be used to derive equations for the kinematics of particles. We have shown that the differential of the displacement with respect to time is equal to the velocity

$$\frac{ds}{dt} = v \tag{8.20}$$

Integrating both sides of the equation with respect to time gives

$$\int \left(\frac{ds}{dt}\right) dt = \int v \, dt$$

that is

$$\int ds = \int v \, dt$$

If $s = s_1$ and s_2 when $t = t_1$ and t_2, respectively, integrating between the appropriate limits leads to

$$\int_{s_1}^{s_2} ds = \int_{t_1}^{t_2} v \, dt \tag{8.21}$$

At first glance, the integral on the left-hand side looks different from those we have dealt with up to now, so we will look at it first. If we consider the function of s,

$$f(s) = cs^n$$

with $c = 1$ and $n = 0$, bearing in mind that $s^0 = 1$, we have

$$f(s) = 1 \times s^0 = 1$$

Multiplying any equation by 1 leaves it unchanged, so

$$\int_{s_1}^{s_2} ds = \int_{s_1}^{s_2} f(s) \, ds = \int_{s_1}^{s_2} 1 \times s^0 \, ds$$

which can be integrated by the rule expressed in Equation 8.15 to give

$$\int_{s_1}^{s_2} 1 \times s^0 \, ds = \left[\frac{1 \times s^{0+1}}{0+1}\right]_{s_1}^{s_2} = \left[\frac{s}{1}\right]_{s_1}^{s_1}$$

$$= s_2 - s_1$$

Now that we have demonstrated this result it is unnecessary to derive it again. Every time we encounter an integral of this type in the future, we shall simply replace it by the difference between the limits,

$$\int_{x_1}^{x_2} dx = x_2 - x_1$$

Returning to Equation 8.21, since the integral $\int_{t_1}^{t_2} v \, dt$ is simply the area under the v–t curve, A_{v-t}, we can write

$$s_2 - s_1 = A_{v-t} \tag{8.22}$$

which is the same as Equation 7.8.

8.5 Constant acceleration

So far we have not put any restriction on how the velocity changes with time. The derivation of Equation 8.22 rests entirely on the statement that velocity is the rate of displacement with time, as expressed by Equation 8.20. If we now require the acceleration to remain constant, that immediately fixes the forms of the a–t, v–t and s–t curves (Figure 7.5) and we can proceed to derive the equations for constant acceleration.

Since the acceleration is given by

$$\frac{dv}{dt} = a$$

integrating with respect to time leads to

$$\int \left(\frac{dv}{dt}\right) dt = \int a \, dt$$

or

$$\int_{v_1}^{v_2} dv = \int_{t_1}^{t_2} a \, dt = \int_{t_1}^{t_2} at^0 \, dt$$

Since the acceleration is constant, a is just a number, and we can write

$$v_2 - v_1 = \left[\frac{at^{0+1}}{0+1}\right]_{t_1}^{t_2} = at_2 - at_1$$

If we choose our datum such that we start counting distance and time from s_1 and t_1 respectively, then $s_1 = 0$ and $t_1 = 0$ and we can now write

$$v_2 = v_1 + at_2$$

which is exactly the same as Equation 7.17. This equation says that the velocity v_2 at time $t = t_2$ equals the initial velocity, v_1 (the velocity when $t = t_1$) plus the product of the acceleration and the time interval. We could write this more generally by saying that the velocity v at any time t is given by

$$v = v_1 + at \tag{8.23}$$

This is actually the equation of a straight line, as can be seen by comparison with Equation 8.1; both equations show the velocity varying linearly with t when the acceleration is constant.

Knowing this property of the velocity when the acceleration is constant allows us to use Equation 8.23 to develop Equation 8.21 still further. Substituting for v in

$$\int_{s_1}^{s_2} ds = \int_{t_1}^{t_2} v \, dt$$

leads to

$$\int_{s_1}^{s_2} ds = \int_{t_1}^{t_2} (v_1 + at) \, dt = \int_{t_1}^{t_2} (v_1 t^0 + at^1) \, dt$$

that is

$$s_2 - s_1 = \left[\frac{v_1 t^1}{1} + \frac{at^2}{2}\right]_{t_1}^{t_2}$$

Introducing the limits with s_1 and $t_1 = 0$ gives

$$s_2 = v_1 t_2 + \frac{1}{2} a t_2^2$$

which is identical to Equation 7.19.

We will now look at one more example of the use of integration for deriving equations of motion. Again we will start with the acceleration and assume it remains constant. We know that

$$\frac{dv}{dt} = a$$

And since $ds/ds = 1$, we have

$$\frac{ds}{ds}\frac{dv}{dt} = a$$

or

$$v\frac{dv}{ds} = a$$

Integrating with respect to s leads to

$$\int v\left(\frac{dv}{ds}\right) ds = \int a \, ds$$

Introduction of appropriate limits produces

$$\int_{v_1}^{v_2} v^1 \, dv = \int_{s_1}^{s_2} as^0 \, ds$$

Therefore

$$\left[\frac{v^2}{2}\right]_{v_1}^{v_2} = \left[\frac{as^1}{1}\right]_{s_1}^{s_2}$$

that is

$$\frac{1}{2}v_2^2 - \frac{1}{2}v_1^2 = as_2$$

and

$$v_2^2 = v_1^2 + 2as_2$$

which, remembering $s_1 = 0$, is the same as Equation 7.18.

We have now derived three of the four equations of motion with constant acceleration that we derived previously, but making use of differentiation and integration. The fourth can be simply derived by eliminating the acceleration from the first two, so we will not pursue it here. Differentiation and integration together form a branch of mathematics known as calculus. Equations of motion can be derived without reference to calculus, but because they are so bound up with the slopes and areas on the s–t, v–t and a–t curves, calculus proves a powerful tool. As problems become more complex, calculus becomes increasingly useful. It is important to understand the use of calculus in the context of the simple problems we have been considering. Even though it is not essential here, an understanding of calculus can be developed further when more difficult problems are encountered.

Recap

- Information relating to displacement, velocity and acceleration can be presented as discrete data in the form (s, t), (v, t) and (a, t), or as discrete points plotted on a graph. It can also be presented in a continuous form by an equation such as $s = f(t)$ or by a continuous curve plotted on a graph. Information derived from experiments is usually in the form of discrete data, whereas information derived from theory is usually in the form of continuous data.

- Continuous data in the form of an equation can be represented graphically by a curve on a graph. The slope at any point on the curve can be determined exactly by differentiating the equation. The area under the curve between a lower and an upper bound can be found exactly by integrating the equation between these limits.

- Continuous data can be approximated from discrete data. Corresponding slopes can be found approximately by numerical differentiation (finite differences) and corresponding areas can be found approximately by numerical integration. Both can be approximated by graphical methods.

- The exact (or approximate) velocity and acceleration can be found as a function of time through the successive theoretical (or numerical) differentiation of the displacement–time and the velocity–time curves respectively.

- Exact (or approximate) changes in the velocity and the displacement over an interval of time can be found by the successive theoretical (or numerical) integration of the acceleration–time and the velocity–time curves over that period.

8.6 Problems

1. A car was travelling at $15\,\mathrm{m\,s^{-1}}$ when its engine was switched off. The distance travelled after this moment was measured every 2 s as the car slowed down and was recorded as follows: $(t, s) = (0, 0, 0)$, (2, 29.90), (4, 59.20), (6, 87.30), (8, 113.60), (10, 137.50), (12, 158.40), (14, 175.70), (16, 188.80), (18, 197.10) and (20, 200.00). Using the formulae $v = \Delta s/\Delta t$ and $a = \Delta v/\Delta t$, draw the s–t, v–t and a–t curves.

2. The motion of the car in the previous example can be represented by the equation $s = -0.0125t^3 + 15t$. By successively taking its derivative with respect to time, t, find expressions for the velocity, v, and the acceleration, a. Compare values of (t, s), (t, v) and (t, a) with those obtained above.

3. A high speed vehicle accelerates from rest as follows: $(t, a) = (0, 3.00)$, (5, 2.53), (10, 2.10), (15, 1.73), (20, 1.40), (25, 1.13), (30, 0.90), (35, 0.73), (40, 0.60), (45, 0.53), (50, 0.50), where the acceleration is measured in $\mathrm{m\,s^{-2}}$. Using numerical integration, estimate how the velocity changes with time over the first 50 s. Hint: if at time t_n the acceleration of the vehicle is a_n and its velocity is v_n, then at time t_{n+1} the velocity will be given by $v_{n+1} = v_n + 0.5(a_n + a_{n+1})(t_{n+1} - t_n)$. Draw the a–t and the v–t curves. Using the same technique, calculate how far the vehicle travels in 10 s.

4. The acceleration of the vehicle in the previous question can be described by the equation $a = t^2/1000 - t/10 + 2.5$. Using calculus derive expressions showing how the velocity, v, and the distance travelled, s, change with time t. Use these expressions to assess the accuracy of the numerical technique used in the previous question.

5. During subsea operations, for a short period of time, the depth, z, of a diving bell beneath the surface of the sea is defined by the equation $z = t^3/3 - 2t^2 - 6$. Plot the v–t curve for the bell during the first 4 s and calculate its acceleration for $t = 0$, 2 and 4 s.

6. The change in velocity, measured in $\mathrm{m\,s^{-1}}$, with time of the sliding part of a machine is described by the equation $v = 0.5t - 1$. Given that the displacement, s, of the part is 0.5 m when $t = 0$, draw the s–t curve for the first 4 s of motion. What is the total *distance*, d, moved by the part in the first 3 s?

7. A particle, starting from rest with zero displacement, moves with an acceleration that varies linearly with time. If, after 6 s, its displacement, s, is 36 m and its acceleration, a, is $4\,\mathrm{m\,s^{-2}}$, what is its velocity, v?

9   Other types of motion

When particles follow paths other than straight lines, they describe curves of varying complexities in two or three dimensions. Then the displacement of the particle must be defined with reference to a coordinate system characterized by two or three variables respectively. Similarly, the velocities and accelerations of the particles executing these motions have two or three components. When the motion is reasonably simple it may be possible to derive equations that describe it. Besides rectilinear motion, the planar curvilinear type is one of the simplest to describe and is the next one we shall consider.

9.1   Planar curvilinear motion

In planar curvilinear motion the particle describes a curve as it travels, but the motion occurs entirely within a single plane. A car travelling along a straight length of motorway that passes through undulating countryside will be describing curvilinear motion as it travels up hill and down dale. All the motion in this example occurs in the vertical plane that includes the line of the road.

The equations for curvilinear motion can be expressed in terms of variables relating to rectangular or polar coordinate systems, or in terms of path variables, as discussed in Section 7.1. Since the motion is two-dimensional, two variables are required to define completely any quantity of interest within whichever coordinate system is adopted. In what follows it is convenient to use path variables. In this system all measurements are made relative to the path followed by the particle. At any instant the distance s travelled by the particle is measured along the path from some fixed point, A, as in Figure 9.1 which shows a particle travelling along the curvilinear path AB.

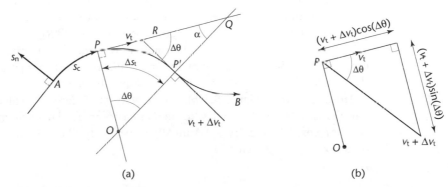

(a)   (b)

Figure 9.1   (a) A particle travelling along a curvilinear path AB. (b) The change in velocity can be resolved into components parallel and at right angles to the tangent at P.

The parameter s is a scalar quantity that measures the magnitude of the displacement, but it provides no information about the particle's direction of travel. Changes in displacement and velocity are measured in directions tangential and normal to the path travelled by the particle at any given instant.

As the car in our illustration travels along the road, unless it is being driven too fast, it will remain in contact with the road, and any given reference point on the car will execute curvilinear motion along a path parallel to the road. As the car moves forward, the distance along the path, s, will continually increase in a direction that is tangential to the path at any given instant in time. The distance between the reference point and the road surface, on the other hand, will always be the same, assuming a rigid suspension! That is to say, the reference point itself always travels along the path, and the displacement of the particle in a direction normal to the path is constant and equal to zero.

Referring to Figure 9.1, we will now consider a particle travelling along the curvilinear path AB. Suppose the particle arrives at the point P at time t, and during the interval Δt the displacement increases by a distance Δs. The average speed of the particle along the path is therefore $\Delta s / \Delta t$, and the instantaneous velocity is given by

$$v_t = \underset{\Delta t \to 0}{\text{limit}} \frac{\Delta s_n}{\Delta t} = \frac{ds}{dt} \quad \text{or} \quad \dot{s}$$

where v_t is directed along the tangent to the path at P. There is no normal velocity component since

$$v_n = \underset{\Delta t \to 0}{\text{limit}} \frac{\Delta s_n}{\Delta t} = \frac{0}{\Delta t} = 0$$

where s_n signifies the displacement normal to the path.

We have seen for rectilinear motion that the acceleration is equal to the rate of change of velocity with respect to time. We will now examine changes in the velocity in directions tangential and normal to the path, in order to derive equations for the tangential and normal components of acceleration.

In Figure 9.1(a) the lines OP and OP' are perpendiculars to the tangents at P and P', therefore the triangles OPQ and $RP'Q$ contain a common angle, α, at Q and right angles at P and P' respectively. The triangles must therefore be similar, and if the angle $\widehat{POQ}$ is equal to $\Delta\theta$ degrees, then so is the angle $\widehat{P'RQ}$, which is also the angle between the two triangles. In other words, over the time interval Δt the particle has changed its direction of travel by the angle $\Delta\theta$. And over that period the magnitude of the velocity has changed; at P' it is now equal to $v_t + \Delta t$.

Turning now to Figure 9.1(b), the velocity at P' can be resolved into components parallel and at right angles to the tangent at P. The velocity components are $(v_t + \Delta t)\cos(\Delta\theta)$ and $(v_t + \Delta t)\sin(\Delta\theta)$ respectively. The change in velocity in the t direction at P is therefore

$$\Delta v_t = (v_t + \Delta v_t)\cos(\Delta\theta) - v_t$$

and the tangential acceleration a_t is given by

$$a_t = \underset{\Delta t \to 0}{\text{limit}} \frac{(v_t + \Delta v_t)\cos(\Delta\theta) - v_t}{\Delta t}$$

At very small angles $\cos(\Delta\theta)$ approaches 1 and the term $v_t \cos(\Delta\theta)$ cancels with the v_t term, leaving

$$a_t = \underset{\Delta t \to 0}{\text{limit}} \frac{\Delta v_t \times 1}{\Delta t}$$

$$a_t = \frac{dv_t}{dt} \quad \text{or} \quad \dot{v}_t \quad \text{or} \quad \ddot{s}_t$$

So far the terms for the velocity and acceleration appear to take a familiar form. However, in curvilinear motion the fact that velocity is a vector quantity assumes a significance it does not have in rectilinear motion. Even if the speed of the particle remains constant, the velocity still changes with time because the direction of travel is continually changing. Because the particle is changing direction there is a component of acceleration normal to the direction of travel as well as tangential to it.

Figure 9.1(b) shows the change in the normal component of velocity Δv_n at P due to the change in direction. This is given by

$$\Delta v_n = (v_t + \Delta v_t)\sin(\Delta\theta) - 0$$
$$= v_t \Delta v_t + \Delta v_t \Delta\theta$$

since $\sin(\Delta\theta) \to \Delta\theta$ for small angles measured in radians.

The normal component of acceleration a_n may therefore be written as

$$a_n = \underset{\Delta t \to 0}{\text{limit}} \frac{\Delta v_n}{\Delta t}$$
$$= \underset{\Delta t \to 0}{\text{limit}} \frac{v_t \Delta\theta + \Delta v_t \Delta\theta}{\Delta t}$$

Since Δv_t and $\Delta\theta$ are both very small quantities, their product is much smaller than $v_t \Delta\theta$, and can safely be considered insignificant. The expression for the normal acceleration then becomes

$$a_n = \underset{\Delta t \to 0}{\text{limit}}\, v_t \frac{\Delta\theta}{\Delta t} = v_t \frac{d\theta}{dt} \quad \text{or} \quad v_t\dot{\theta} \tag{9.1}$$

The term $\dot{\theta}$ is the rate of change of direction of the particle with time, and is known as the angular velocity. The angular velocity is measured in radians per second.

It is not always convenient to deal with curvilinear motion directly in terms of the angular velocity of the particle, and it is useful to derive other equations for the normal component of acceleration. This requires the ability to calculate the distance s measured along the arc of a circle of radius r subtended by the angle θ at its centre.

If a radial line of constant length r fixed at the point O, as shown in Figure 9.2, rotates through an angle of θ radians, its free end describes a circular arc of length s. If it completes a whole revolution, it rotates through 2π radians (360°) and the arc length s equals the circumference $2\pi r$. The arc length s is the same fraction of the circumference $2\pi r$ as the fraction θ (in radians) of one complete revolution (2π

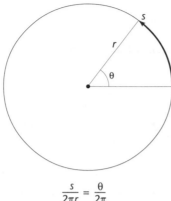

$$\frac{s}{2\pi r} = \frac{\theta}{2\pi}$$

Figure 9.2 Relationship between arc and angle.

radians), so if the line rotates through $\frac{1}{4}$ of a revolution, $\theta = \pi/2$ radians (90°) and the arc length $s = \frac{1}{4}$ of the complete circumference, therefore

$$s = \frac{\pi/2}{2\pi}(2\pi r) = \frac{1}{4}(2\pi r)$$

$$\frac{s}{2\pi r} = \frac{\theta}{2\pi}$$

$$s = \left(\frac{\theta}{2\pi}\right)2\pi r \qquad (9.2)$$

Simplifying Equation 9.2 leads to

$$s = r\theta$$

where θ is measured in radians.

Returning now to Figure 9.1, if $\Delta\theta$ is very small the length of the arc Δs between P and P', which is the distance travelled during the time interval Δt, can be written as

$$\Delta s = OP\,\Delta\theta \qquad (9.3)$$

Also, as $\Delta\theta$, and so Δs, becomes very small, the lengths OP and OP' tend towards the same value ρ, that is

$$OP' = \underset{\Delta t \to 0}{\text{limit }} OP = \rho$$

where ρ is known as the radius of curvature of the curve at the point P. The point O is called the centre of curvature. The normal component of acceleration is always directed towards the centre of curvature.

Rewriting Equation 9.3 gives

$$\frac{\Delta\theta}{\Delta s} = \frac{1}{OP}$$

and in the limit, as Δs approaches zero, we have

$$\frac{d\theta}{ds} = \underset{\Delta t \to 0}{\text{limit }} \frac{\Delta\theta}{\Delta s} = \underset{\Delta t \to 0}{\text{limit }} \frac{1}{OP} = \frac{1}{\rho} \qquad (9.4)$$

As we have seen, the normal component of the acceleration can be expressed as

$$a_n = v_t \frac{d\theta}{dt}$$

$$= v_t \frac{ds}{ds} \frac{d\theta}{dt}$$

$$= v_t \frac{ds}{dt} \frac{d\theta}{ds} = v_t v_t \left(\frac{1}{\rho}\right)$$

$$a_n = \frac{1}{\rho} v_t^2 \qquad (9.5)$$

Equation 9.4 can also be used to express the tangential velocity, v_t, in terms of the angular velocity $\dot{\theta}$. Since

$$\frac{d\theta}{ds} = \frac{d\theta}{ds} \frac{dt}{dt} = \frac{d\theta}{dt} \frac{dt}{ds} = \frac{1}{\rho}$$

$$\dot{\theta}\left(\frac{1}{v_t}\right) = \frac{1}{\rho}$$

$$v_t = \rho\dot{\theta}$$

Combining this equation with Equation 9.5 allows the normal component of acceleration to be expressed also in terms of the angular velocity $\dot{\theta}$:

$$a_n = v_t\dot{\theta} = \rho\dot{\theta}^2$$

We have now derived several equations describing curvilinear motion, and it is time for a recap.

Recap

■ The equations for curvilinear motion may be expressed in terms of path variables in which a measure of the displacement is the scalar distance s travelled by the particle along the path relative to a fixed point on the path.

■ The velocity v_t at any given time is the rate of change of s with respect to time, and its direction is tangential to the particle's path at that time. There is no component of velocity normal to the path.

■ The tangential velocity can be written

$$v_t = \frac{ds}{dt} = \dot{s} \quad \text{or} \quad v_t = \rho\frac{d\theta}{dt} = \rho\dot{\theta} \qquad (9.6)$$

■ The acceleration has two components, a tangential component a_t directed along the path, and a normal component a_n directed towards the centre of curvature. The magnitude of the acceleration is given by the vector sum of a_t and a_n:

$$|a| = (a_t^2 + a_n^2)^{1/2} \qquad (9.7)$$

where a is directed at an angle of α away from the tangent and is given by

$$\alpha = \tan^{-1}\left(\frac{a_n}{a_t}\right) \tag{9.8}$$

as shown in Figure 9.3.

■ The tangential component of acceleration can be written

$$a_t = \frac{dv_t}{dt} = \dot{v}_t = \ddot{s} \quad \text{or} \quad a_t = \rho\ddot{\theta} \tag{9.9}$$

■ The normal component of acceleration can be written

$$a_n = v_t\dot{\theta} \quad \text{or} \quad a_n = \frac{1}{\rho}v_t^2 \quad \text{or} \quad a_n = \rho\dot{\theta}^2 \tag{9.10}$$

➤ **Example 9.1**

Tests are being carried out to measure the performance of a prototype car for a new design. The car is equipped with instruments that, among other things, register the speed along the road and the rate of change of that speed with time. The tests are being carried out on a track whose bends are of known radius of curvature. At a particular instant, as the car rounds a bend of radius of curvature 400 m, the instruments record a speed of 140 kph and a rate of change of speed of 10 kph per second. What is the magnitude and direction of the total acceleration of the car at that moment?

Solution If the track is level, the car is executing plane curvilinear motion. It will possess a velocity v_t whose magnitude is equal to 140 kph and whose direction is tangential to the bend. Because its speed is changing, it will also experience a tangential component of acceleration a_t whose magnitude is equal to 10 kph per second.

The first step in finding the car's total acceleration is to convert these quantities to a consistent set of units.

$$v_t = 140 \times \frac{1000}{60 \times 60} = 38.89 \text{ m s}^{-1}$$

$$a_t = 10 \times \frac{1000}{60 \times 60 \times 1} = 2.78 \text{ m s}^{-2}$$

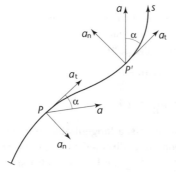

Figure 9.3 The magnitude of an acceleration is the vector sum of its normal and tangential components.

In curvilinear motion the acceleration can be expressed in terms of two components, the tangential acceleration a_t, which is known, and the normal acceleration, a_n, which is not. But Equation 9.10 shows that the normal acceleration can be calculated if the tangential velocity v_t and the radius of curvature ρ are known:

$$a_n = \frac{1}{\rho} v_t^2 = \frac{1}{400} \times (38.89)^2 = 3.78 \text{ m s}^{-2}$$

Acceleration is a vector quantity, so it has direction as well as magnitude. The component a_t is directed along the tangent to the curve. The component a_n acts at right angles to a_t and is directed towards the centre of curvature – the inside of the bend. The total acceleration is the resultant of these two components and its magnitude $|a|$ and direction α can be calculated from Equations 9.7 and 9.8 respectively.

$$|a| = (a_t^2 + a_n^2)^{1/2} = (2.78^2 + 3.78^2)^{1/2} = 4.69 \text{ m s}^{-2}$$

$$\alpha = \tan^{-1}\left(\frac{a_n}{a_t}\right) = \tan^{-1}\left(\frac{3.78}{2.78}\right) = 53.67°$$

The magnitude of the total acceleration a is **4.68 m s^{-2}** and it is directed towards the inside of the bend at an angle of **53.67°** to the tangent to the bend.

9.2 Circular motion

Circular motion is a special case of plane curvilinear motion in which the radius of curvature remains constant and equal to the radius of a circle r_c, that is $\rho = r_c$. The equations describing the special case of circular motion are the same as those just derived for the general case of plane curvilinear motion except that now

$$v_t = r_c \frac{d\theta}{dt} = r_c \dot{\theta} \tag{9.11}$$

$$a_t = r_c \ddot{\theta} \tag{9.12}$$

and

$$a_n = \frac{1}{r_c} v_t^2 \quad \text{or} \quad a_n = r_c \dot{\theta}^2 \tag{9.13}$$

All the other equations derived earlier remain the same.

Example 9.2

Suppose that the prototype car of Example 9.1 is being tested on a perfectly circular track of diameter 1 km and that, among other quantities, the instrumentation is designed to measure the speed and the total acceleration with readouts of the two components a_t and a_n. If the equipment malfunctions and, at a given moment, it is only possible to obtain the two readings $|a| = 5.0$ m s^{-2} and $a_t = 2.5$ m s^{-2}, what speed is the car doing at that moment?

Solution The total acceleration a, whose magnitude is known, is the vector sum of a_t, which is also known, and a_n. Since two of the three quantities are known, a_n can be found from Equation 9.7, which connects them all together.

$$(a_t^2 + a_n^2)^{1/2} = (2.5^2 + a_n^2)^{1/2} = 5.0 \text{ m s}^{-2}$$

$$a_n = \sqrt{5^2 - 2.5^2} = 4.33 \text{ m s}^{-2}$$

Knowing a_n and r_c, the radius of the track, allows the calculation of the speed, v_t, from Equation 9.13.

$$a_n = \frac{1}{r_c} v_t^2$$

$$v_t = \sqrt{a_n \times r_c} = 46.53 \text{ m s}^{-1}$$

The car is travelling at 46.53 m s^{-1} at the moment the accelerations were measured.

9.3 Periodic motion

If a particle is travelling along a circular path at constant speed, it will take the same time T to complete a circuit on each revolution. Every T seconds the particle will describe a path identical to the one it followed in the previous T seconds. Circular motion at constant speed is just one example of a type of motion called **periodic motion** in which the time taken to complete one cycle of a regular repeated sequence of displacements is fixed and is known as the **period, T**.

In the case of periodic circular motion, the frequency f with which the particle passes any given reference point on its path is given by

$$f = \frac{1}{T} \quad \text{(Hz)} \tag{9.14}$$

For example, if the particle takes 0.5 s to complete a circuit ($T = 0.5 \text{ s}$) it passes the reference point twice in every second. That is, the characteristic frequency of the motion, f, is 2 cycles per second, or 2 Hz, where the symbol Hz has been adopted as the unit of frequency in recognition of the famous scientist Heinrich Hertz.

If the angle between the reference point and the instantaneous position of the particle subtended at the centre of the circle is θ, then θ increases from 0 to 360°, or from 0 to 2π radians, each time the particle executes a complete circuit in time T. The constant rate of change of angle with time is called the angular velocity. Measured in radians per second and denoted by the symbol ω, the angular velocity may be found from

$$\omega = \frac{2\pi}{T} \quad \text{(rad s}^{-1}) \tag{9.15}$$

And Equation 9.14 shows that

$$\omega = \frac{2\pi}{T} = \frac{2\pi}{1/f} = 2\pi f \quad \text{(rad s}^{-1}) \tag{9.16}$$

For circular motion with a fixed period, the angular velocity ω equals 2π times the frequency of the motion. It is often convenient to express the frequency of other types of periodic motion as $2\pi f$, when it is called the **circular frequency** or the **angular frequency, ω**.

Periodic motion involves regular repeated sequences of displacement. Typically, a body experiencing periodic motion makes positive and negative excursions about some fixed reference point. Examples might be a chair on a fairground wheel rising above then sinking below its central spindle, or a cork bobbing up and down about the mean water level in regular waves. Several examples of periodic motion are illustrated in Figure 9.4, where the displacements of the bodies involved are plotted against time. Most of these examples occur only in very special circumstances, but a sinusoidal displacement–time (s–t) curve crops up in many engineering applications.

9.4 Harmonic motion

If the variation of the displacement of an object with time can be represented by a sinusoid (sine or cosine curve) it is undergoing **harmonic motion**. As we shall see, it is a characteristic feature of this type of motion that the acceleration of the object is always acting in the opposite direction to its displacement. Consequently, there is always a **restoring force** which tends to move the body back towards its equilibrium position. Motion in which no other forces are acting is called **simple harmonic motion**. Commonly quoted examples of simple harmonic motion are a freely swinging pendulum and a weight oscillating at the end of a spring. In these examples the motion is represented by sinusoids of constant amplitude. No external force is applied (except to provide the initial displacement); gravity and the spring stiffness provide restoring forces which cause the objects to reverse their directions at the extremities of their displacement.

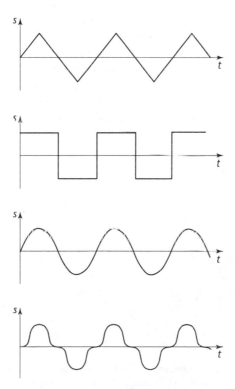

Figure 9.4 Examples of periodic motion.

In reality, a pendulum experiences other forces arising from friction at its point of attachment and air resistance, so the motion gradually dies away. Forces such as these are termed **damping forces**. When there are damping forces, the motion is only sinusoidal if some other agent causes it to be so; this type of motion is called **forced motion**. As we are dealing with kinematics, we are more interested in modelling the motion of the particle than in the forces that cause it, but the motion and the forces are inextricably linked. For convenience, let us look at harmonic motion in the context of forced motion.

The mechanism shown in Figure 9.5 consists of a rotating disc and a horizontal arm. As the disc rotates, a peg located on its rim slides backwards and forwards in a slot in a horizontal arm that is free to move vertically up and down. The horizontal arm undergoes a complete cycle of its motion every time the disc executes a full revolution. A pen attached to the arm records its vertical displacement on a chart recorder, which consists of a paper chart being steadily wound from one drum to another mounted in the vertical plane. The resulting trace on the chart is an s–t curve for the displacement of the arm.

The peak displacement $\hat{s}$ equals the disc radius r_c; the instantaneous displacement, measured from the centre of the disc, is given by

$$s = r_c \sin \theta = \hat{s} \sin \theta \tag{9.17}$$

The trace on the chart recorder shows how the displacement s varies with time. For every instant of time, there is a corresponding value of the angle θ, which is a measure of how much the disc has rotated, or its **angular displacement**. The angular displacement increases through 2π radians for every revolution of the disc.

Below the s–t curve is plotted the corresponding change of displacement of the arm with θ, i.e. the s–θ curve whose equation is given by Equation 9.17. The displacements on the s–t curve with t varying from 0 to T are identical to those on the s–θ curve with θ varying from 0 to 2π radians. At corresponding points on the two curves, where the displacements are equal, the ratio of t to T equals the ratio of θ to 2π, so

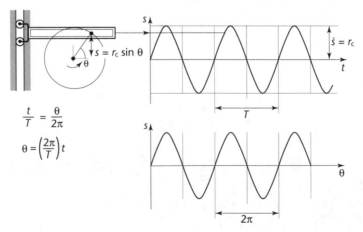

$$\frac{t}{T} = \frac{\theta}{2\pi}$$

$$\theta = \left(\frac{2\pi}{T}\right)t$$

Figure 9.5 A rotating disc and a horizontal arm. The horizontal arm is free to move in the vertical direction and carries a pen which marks a chart wound at a constant speed from a revolving drum. Typical traces are shown on the right.

$$\frac{t}{T} = \frac{\theta}{2\pi}$$

$$\theta = \left(\frac{2\pi}{T}\right)t = \omega t \qquad (9.18)$$

This shows there is a one-to-one relationship between θ and t. For example, halfway through the first cycle of the s–t curve $t = T/2$ The corresponding point on the s–θ curve is given by

$$\theta = \omega t = \frac{2\pi}{T} \times \frac{T}{2} = \pi \quad \text{(radians)}$$

which is halfway through the cycle. The displacement is the same and equal to zero on both curves.

Substituting for θ from Equation 9.18 into Equation 9.17 leads to the equation for the s–t curve written in terms of the circular frequency of the disc as

$$s = \hat{s} \sin \theta = \hat{s} \sin \omega t \qquad (9.19)$$

An s–θ curve can be derived for all harmonic motion, whether or not it is generated by a mechanism involving a circular disc, by equating the displacements occurring over a period of 2π radians with those occurring over the period of T seconds of the motion. Similarly, harmonic motion can always be expressed in terms of a corresponding circular frequency ω and the time t as given by Equations 9.18 and 9.19.

Since the displacement can be expressed as a function of time, it is possible to find the equations for the velocity and acceleration by successive differentiation. The velocity is given by differentiating the displacement with respect to time:

$$\frac{ds}{dt} = \dot{s} = \omega\hat{s} \cos \omega t \qquad (9.20)$$

Similarly, the acceleration is given by differentiating the velocity with respect to time:

$$\frac{d^2s}{dt^2} = \ddot{s} = -\omega^2\hat{y} \sin \omega t \qquad (9.21)$$

In Figure 9.6 the displacement, velocity and acceleration are plotted against the angular displacement, $\theta = \omega t$, where

$$s = \hat{s} \sin \omega t = \hat{s} \sin \theta$$

$$\dot{s} = \omega\hat{s} \cos \omega t = \omega\hat{s} \cos \theta$$

$$\ddot{s} = -\omega^2\hat{s} \sin \omega t = -\omega^2\hat{s} \sin \theta$$

The three quantities are all sinusoids but they have different amplitudes. If the peaks and troughs of the curves had all occurred at corresponding values of θ, the quantities would have been **in phase** with each other. However, this is not the case and they are not in phase.

There is a standard trigonometrical identity (see Further Reading) which says

$$\sin(\theta + \varepsilon) = \sin \theta \cos \varepsilon + \cos \theta \sin \varepsilon$$

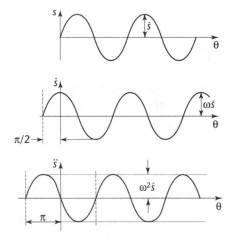

Figure 9.6　Displacement, velocity and acceleration plotted against angular displacement for a simple harmonic motion.

Therefore

$$\sin(\theta + \pi) = \sin \theta(-1) + \cos \theta\,(0) = -\sin \theta$$

$$\sin\left(\theta + \frac{\pi}{2}\right) = \sin \theta\,(0) + \cos \theta\,(1) = \cos \theta$$

Therefore the velocity can be written

$$\dot{s} = \omega\hat{s}\,\sin(\theta + \varepsilon)\qquad \text{where } \varepsilon = \pi/2$$

and, similarly, the acceleration can be written

$$\ddot{s} = \omega^2\hat{s}\,\sin(\theta + \varepsilon)\qquad \text{where } \varepsilon = \pi$$

The angle ε is known as the **phase angle** or the **phase shift** between the displacement and the velocity ($\varepsilon = \pi/2$) or the displacement and the acceleration ($\varepsilon = \pi$). In this case the velocity $\dot{s}$ **leads** the displacement s by $\pi/2$ radians, and the acceleration $\ddot{s}$ **leads** or **lags** the displacement s by π radians. That is, the acceleration is π radians, or 180°, **out of phase** with the displacement.

If t_ε is the time shift corresponding to the phase angle ε, it follows from Equation 9.17 that

$$\varepsilon = \left(\frac{2\pi}{T}\right)t_\varepsilon = \omega t_\varepsilon$$

So the velocity and acceleration can also be written

$$\dot{s} = \omega\hat{s}\,\sin \omega(t + t_\varepsilon)\quad \text{where } t_\varepsilon = T/4$$

and

$$\ddot{s} = \omega^2\hat{s}\,\sin \omega(t + t_\varepsilon)\quad \text{where } t_\varepsilon = T/2$$

The mechanism shown in Figure 9.5 forced the horizontal arm to move in such a way that its vertical motion could be represented by a sine wave of constant amplitude. A body moving freely with this type of motion, with only a restoring force acting on it, is undergoing simple harmonic motion. All the equations that have just been deduced from the behaviour of the mechanism illustrated in Figure 9.5 apply equally well to simple harmonic motion.

Recap

▪ Simple harmonic motion is a form of periodic motion whose displacement can be represented by a sinusoid of constant amplitude $\hat{s}$ and period T.

▪ If the frequency of motion is f Hz (cycles per second), its circular frequency ω and angular displacement θ are given by

$$\omega = 2\pi f = \frac{2\pi}{T} \quad (\text{rad s}^{-1})$$

$$\text{and} \quad \theta = \omega t \quad (\text{rad}) \tag{9.22}$$

▪ The equation for the displacement is

$$s = \hat{s} \sin \theta = \hat{s} \sin \omega t \tag{9.23}$$

▪ The velocity can be obtained by differentiating the displacement with respect to time, and can also be expressed in terms of a sine curve with a phase angle $\varepsilon = \pi/2$:

$$\dot{s} = \omega\hat{s} \cos \theta = \omega\hat{s} \cos \omega t$$

$$= \omega\hat{s} \sin(\theta + \varepsilon) \quad \varepsilon = \frac{\pi}{2} \tag{9.24}$$

▪ Similarly the acceleration can be obtained by differentiating the velocity; it can also be written in terms of a sine curve, but with a phase angle $\varepsilon = \pi$:

$$\ddot{s} = -\omega^2\hat{s} \sin \theta = -\omega^2\hat{s} \sin \omega t$$

$$= \omega^2\hat{s} \sin(\theta + \varepsilon) \quad \varepsilon = \pi \tag{9.25}$$

➤ **Example 9.3**

A weight suspended from a light frictionless spring is oscillating about a mean position with simple harmonic motion of amplitude 10 cm and period 10 s. Calculate its circular frequency ω and the maximum velocity and acceleration it achieves.

Solution The circular frequency of the motion can be calculated immediately from Equation 9.21.

$$\omega = \frac{2\pi}{T} = \frac{2\pi}{10}$$

$$= 0.628 \text{ rad s}^{-1}$$

If the amplitude of the motion is 10 cm, $\hat{s} = 0.1$ m. The velocity is given by Equation 9.24.

$$\dot{s} = \omega\hat{s} \sin\left(\theta + \frac{\pi}{2}\right) \tag{9.26}$$

Since the maximum value that can be taken by $\sin(\theta + \varepsilon)$ is 1, the maximum velocity is given by

$$\dot{s}_{max} = \omega\hat{s} \times 1 = 0.628 \times 0.1 \times 1$$
$$= 0.0628 \text{ m s}^{-1}$$

Similarly, the acceleration can also be expressed as a sine curve as given by Equation 9.25, so the maximum acceleration is

$$\ddot{s}_{max} = \omega^2\hat{s} \times 1 = 0.628^2 \times 0.1 \times 1$$
$$= 0.0394 \text{ m s}^{-2}$$

The circular frequency of the motion is **0.628 rad s^{-1}**, the maximum velocity is **0.0628 m s^{-1}** and the maximum acceleration is **0.0394 m s^{-2}**.

9.5 Problems

1. A flywheel of 1 m diameter is spinning with a constant angular velocity of 150 rad s^{-1}. Calculate the velocity and the acceleration of a point located on its rim.

2. A car with wheels of diameter 0.55 m is travelling at 80 kph. Calculate the velocity and acceleration relative to the centre of the wheel hub of a small stone wedged in the tread of one of the tyres. If the car comes to rest with uniform deceleration in 173 m, what is the angular deceleration of the wheel and how many revolutions does it experience before the car stops?

3. A go-kart travelling at constant speed rounds a horizontal bend whose radius of curvature is 50 m. If the go-kart starts to slide when the normal acceleration, a_n, exceeds 0.8 g, calculate the maximum speed it can travel without sliding. (g is the acceleration due to gravity $= 9.81$ m s^{-2}.

4. A ship is making a turning manoeuvre in which its heading changes at a constant rate. If it is travelling at 10 m s^{-1} and its heading changes by 1° every second, calculate its acceleration.

5. A train travelling round a gradual bend increases its speed at a constant rate of 4.5 m s^{-1} every minute. Calculate its speed at a point on the curve where the radius of curvature is 1000 m if the total acceleration of the train is 0.903 m s^{-2}.

6. A chair on a fairground big wheel has an angular velocity, relative to the centre of the wheel, of 0.1 rad s^{-1} and its speed is decreasing at a rate of 1.25 m s^{-2}. Calculate the radius of the wheel if the chair always remains vertically upright and its total acceleration is 1.275 m s^{-2}.

7. A weight suspended from a light elastic spring is oscillating about a reference point with a constant period, T, and amplitude, $\hat{s}$, of 2 s and 0.05 m respectively. Estimate the circular frequency, ω, and the maximum magnitude of the velocity and the acceleration.

8. The maximum vertical velocity and acceleration of a waverider buoy as it floats up and down in regular waves are π m s^{-1} and $\pi^2/10$ m s^{-2} respectively. Estimate the amplitude and period of the waves.

Kinetics

Part III is devoted to kinetics, which studies the relationship between the motions of bodies and the forces that cause them. This is the second part of the study of dynamics, the branch of mechanics concerned with the motions of bodies under the action of out-of-balance forces.

10 Forces and motion

Part II is concerned with kinematics, the study of motion without reference to the forces causing it. Part III is devoted to **kinetics** which studies forces acting on a body and how they produce motion. Chapter 3 introduced Newton's laws of motion and the first and third laws were used as the basis for statics. Statics looks at situations in which the forces acting on the body are balanced, or sum to zero. Consequently, and as stated by Newton's laws, the body remains at rest, or continues to move with constant velocity. We are now going to look at situations in which the forces acting on a body are unbalanced.

Newton's second law describes the relationship between the unbalanced forces acting on a body and its acceleration. If we know the unbalanced forces acting on a body, through Newton's second law we can deduce its acceleration and predict its motion. Similarly, it is possible to predict the accelerations experienced by a body from knowledge of its motions. For example, it is possible to calculate the motions of a passenger ferry in a given sea state. Using Newton's second law it is then possible to calculate the accelerations experienced by people and objects in, for example, the restaurant. This procedure can be used in the preliminary design of the ferry to ensure the cutlery and plates remain on the table, or to avoid the passengers becoming seasick (unless they have very weak stomachs). Even more important, the vessel can be designed so it doesn't capsize and its cargo remains secure. There are many other applications in all branches of engineering and science where dynamics plays a crucial role in the safe and efficient design of artifacts used by the world at large.

10.1 Newton's second law revisited

The version of Newton's second law presented in Chapter 3 is a special case of a more general law. It makes the assumption that the mass of the body remains constant. In many applications this assumption is perfectly valid to all intents and purposes. In others its adoption may lead to serious errors. At the start of its mission, for example, a rocket's fuel represents a significant proportion of its mass. But by the end of its mission the mass of the fuel should be insignificant in comparison to the mass of the rocket body. Before proceeding to the more general form of Newton's second law, it is helpful to consider in more detail some of the terms we will use.

The **mass** of a body is a measure of the amount of material from which it is constituted. At its most fundamental, it is a measure of the number of atoms making up the body; it should not be confused with weight. **Weight** is a measure of the force attracting the body towards the centre of the earth due to the earth's gravitational pull.

Weight and mass are often confused in everyday life and this is a matter to which we will return.

Very useful in describing the motion of bodies is the concept of momentum. The **momentum** of a body is simply the product of its mass and velocity. A body of mass m travelling with a velocity v has a momentum of mv. The more general version of Newton's second law is framed in terms of changes in the momentum of a body, and may be stated as follows:

> *The rate of change of momentum of a particle with respect to time is proportional to the resultant force acting on it and is in the direction of this force.*

The law may be written as an equation in the following manner:

$$F = c\frac{d}{dt}(mv) \tag{10.1}$$

where c is a constant of proportionality. If the mass of the body does not change in time, the equation can be developed further as

$$F = cm\frac{dv}{dt} = cma \tag{10.2}$$

Having discussed the concept of mass, let us introduce the related concept of **inertia**. Suppose a particular body is subjected to a given force and the resulting acceleration is measured. When this experiment has been completed, the data relating to the body will consist of a series of forces, F_1, F_2, ..., F_n, and a corresponding set of accelerations, a_1, a_2, ..., a_n. If the forces are divided by the accelerations:

$$\frac{F_1}{a_1} = \frac{F_2}{a_2} = \ldots \frac{F_n}{a_n} = c^* \tag{10.3}$$

the same constant, c^*, is obtained in every case, within experimental accuracy. Many experiments involving many different bodies in many different experimental arrangements have been conducted on this same basic theme, and they all give the same result. The result is consistent with Newton's second law, itself proposed following experimental observation. It also indicates that, for any given experiment, the constant c^* is a measure of some unchanging property of the body involved. That property, its capacity to resist changes in its motion, is known as inertia.

A body having a large inertia will need a large force to give it a large acceleration. By comparing Equations 10.2 and 10.3, notice that the constant c^* is in fact equal to cm, so it is proportional to the mass. In other words, the mass of a body is a quantitative measure of its inertia.

So far we have discussed Newton's second law without reference to units. In the SI units we have been using, the units of mass, length and time are respectively kilograms (kg), metres (m) and seconds (s). The unit of force is the newton (N), partly in recognition of Newton's great contribution to the field of dynamics.

> *A newton is the force required to give a mass of one kilogram an acceleration of one metre per second per second.*

If we substitute 1 N, 1 kg and 1 m s^{-2} for the force, mass and acceleration respectively in Equation 10.2, the only remaining unknown is the constant c. Then

$$1 = c \times 1 \times 1$$

Therefore, when using SI units, the constant of proportionality c has a value of unity and, if the mass is constant, Newton's second law may be written

$$F = ma \tag{10.4}$$

The other reason why the SI unit of force is called the newton is because it is deliberately defined so the constant of proportionality becomes unity and Equation 10.4 can be used.

Before moving on to study the application of Newton's second law, we will use it to consider the relationship between mass and weight. On earth everything is subjected to the earth's gravitational field, whose effect is to attract bodies towards its centre. If a body is only subjected to the force due to the gravitational field, it will move towards the centre of the earth with an acceleration that depends on its distance from the earth's centre. At the surface of the earth, the acceleration due to gravity, usually denoted by the symbol g, equals approximately $9.81\,\mathrm{m\,s^{-2}}$. It can be seen from Newton's second law that a body of mass m must experience a gravitational force directed towards the earth's centre and of magnitude mg. That force is the body's weight.

> **The weight of a body is the force with which it is attracted to the centre of the earth by the earth's gravitational pull.**

Since it is a force, weight is a vector quantity that has both magnitude and direction. The SI unit of weight is the newton. The weight of a given body depends on its position relative to the centre of the earth. The further away a body moves from the earth, the weaker the gravitational pull and the smaller its weight. In this day and age, most people are aware of the weightlessness experienced by astronauts when they escape the earth's gravitational pull; weightlessness is an example of this effect.

When an object rests on the surface of the earth, it is attracted towards the centre of the earth (according to Newton's second law) with a force mg, that is by its weight. However, it experiences a reaction provided by the earth that is equal and opposite to its weight (in accordance with Newton's third law), so it remains stationary relative to the earth (according to Newton's first law).

Mass, on the other hand, is a measure of the amount of matter making up a body and a measure of that body's resistance to changes in its motion, i.e. its inertia. The total mass making up a body remains constant, although its weight may vary. But even mass may be treated as variable, like the rocket burning fuel. Nevertheless, the total mass of the rocket, the unburnt fuel and the burnt fuel does remain constant. It is just convenient to exclude the burnt fuel from the calculations.

Unlike weight, mass is a scalar quantity having only magnitude. In the SI system the unit of mass is the kilogram. Originally a kilogram was defined as the mass of 1000 cubic centimetres of water. For various reasons this definition has become obsolete, and a kilogram is now defined as the mass of a particular cylinder of platinum kept by the International Bureau of Standards near Paris in France.

➤ **Example 10.1**

The gravitational pull of the earth can be calculated from Newton's law of gravitation. Newton's law of gravitation describes the mutual force of attraction F between two bodies of masses m_1 and m_2 separated by a distance r; it can be written as

$$F = c\frac{m_1 m_2}{r^2} \tag{10.5}$$

where c is a universal constant known as the constant of gravitation and which can be taken to equal 6.673 m³ kg⁻¹ s⁻². Calculate the acceleration due to gravity g; calculate the weight of a body of mass 1 kg at the earth's surface and at an altitude of 1000 km above the earth's surface. The earth may be taken as a sphere of mass 5.976×10^{24} kg and radius 6.371×10^3 km.

Solution The force of mutual attraction between the body and the earth is simply the gravitational force acting on the body; so if h is the altitude, we may write

$$F = m_1 g = c\frac{m_1 m_2}{(r+h)^2}$$

$$g = c\frac{m_2}{(r+h)^2}$$

$$= 6.673 \times 10^{-11} \times \frac{5.976 \times 10^{24}}{(6.371 \times 10^3 \times 10^3 + h)^2} \tag{10.6}$$

Substituting $h = 0$ and $h = 1000 \times 1000$ m in Equation 10.6, we get the acceleration due to gravity, $g = 9.825$ m s⁻² and 7.340 m s⁻² respectively.

The weights of the body at the two different altitudes can be found from Newton's second law, with the acceleration equal to the acceleration due to gravity, $F = mg$. Since the mass of the body is 1 kg, its weight is numerically equal to the value of g, although the units are newtons.

On the earth's surface, the acceleration due to gravity is $g = 9.825$ m s⁻² and the weight of the body is $g = $ **9.825 N**. At an altitude of 1000 km the acceleration due to gravity is 7.340 m s⁻² and the weight of the body is **7.340 N**.

Notice how the value for the acceleration due to gravity on the earth's surface is slightly different from the value of 9.81 m s⁻² quoted earlier. The calculation made the assumption that the earth is a perfect sphere of radius 6.371×10^3 km. It also ignored the fact that the earth is spinning. In fact, the earth is flattened towards the poles and the distance from its centre to its surface varies with geographical location. For this reason the acceleration due to gravity at the surface of the earth also varies from place to place. For most engineering calculations these and other factors are ignored and a value of 9.81 m s⁻² is generally taken for the acceleration due to gravity at sea level.

➤ **Example 10.2**

In Figure 10.1(a) a mass of m kg is being raised through the application of a pulley system by a weight whose mass is $2m$ kg. In Figure 10.1(b) the same mass is being raised by the direct application of a force of $2mg$ N to the pulley system. In both cases the mass of the cable in the pulley system is negligible in comparison to the mass of the weights. Both systems have out-of-balance forces applied to them, so the mass will accelerate in both cases. Which system, if any, gives the greater acceleration?

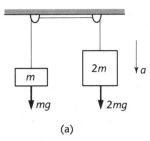 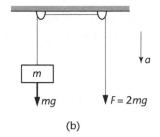

(a)   (b)

Figure 10.1 See Example 10.2.

Solution   In case (a) the weight of the smaller mass, mg, opposes the acceleration whereas the weight of the larger mass, $2mg$, acts in the same direction as the acceleration of the two weights. The net out-of-balance force, F, acting in the direction of the acceleration is therefore given by

$$F = -mg + 2mg$$

The total mass of the system is $3m$ kg. Substituting into Newton's second law gives

$$F = ma$$

$$-mg + 2mg = 3ma$$

$$a = \frac{g}{3} \text{ m s}^{-2}$$

In case (b) the smaller weight again opposes the acceleration and a force of $2mg$ N is applied in the direction of the acceleration in place of the larger weight of case (a). The net out-of-balance force, F, in the direction of the acceleration is therefore the same in both cases. In case (b), however, the total mass of the system is m kg. Substituting these quantities into Newton's second law now leads to

$$F = ma$$

$$-mg + 2mg = ma$$

$$a = g \text{ m s}^{-2}$$

In case (b), therefore, the acceleration is three times greater than in case (a). At first glance, it may have been expected from Figure 10.1 that the acceleration would be the same in both cases. A little reflection shows that the result obtained is consistent with Newton's second law (which means there are no numerical slips) and the concept of inertia. The out-of-balance force is the same in both cases, but the inertia of the first system, which has a mass of $3m$ kg, is much bigger than the inertia of the second system, which has a mass of only m kg. The system with the larger inertia attains the smaller acceleration in response to the same out-of-balance force, as would be expected.

10.2 Approaching problem solving

A knowledge of dynamics leads to an understanding of many naturally occurring phenomena. It is also an essential prerequisite to the safe and efficient design of many items. Dynamics has many applications in engineering, most of them concerned with evaluating the forces to which a particular design element will be subjected, or how the element will respond kinematically to the forces engendered by a particular environment.

Whether you are a scientist trying to explain some natural phenomenon, or an engineer involved in design, your first step in dealing with any problem will be in correctly identifying and defining it. Nearly all objects of interest interact with other objects and forces making up a larger and often very complex system. Before it can be analysed, it is usually necessary to isolate the relevant object from its surroundings and to identify the forces they impose upon it.

In the two examples we have just looked at, we were able to consider the systems as whole systems. Had we set out to calculate some other quantity, this may not have been possible. Suppose, for example, we had been asked to calculate the tension in the cable of the pulley system shown in Figure 10.1. In order to solve this problem, we need to explore how the two weights interact with each other and how both interact with the cable.

The first step is to identify the forces that are operating on the system. Consider the smaller weight in Figure 10.1(a). It is obviously being acted upon by gravity; it is also interacting with the cable, which in turn is interacting with the larger weight, which is also subjected to the force of gravity. The interaction induces a tension T, which is constant throughout the cable since both weights move at the same speed.

Suppose we now cut the cable just above the smaller weight, as shown in Figure 10.2, and at both cut ends, we instantaneously apply a force equal to the tension. The small weight continues to experience the same force, equal to the tension of the cable, as it experienced before the cut. The cable continues to experience the same tension throughout its length. The system as a whole behaves as though the cut did not exist, but now the smaller weight has been isolated from the rest of the system. We can now draw a boundary, marked A in the figure, through the cut and enclosing the weight and the forces acting on it. The boundary contains the **free body diagram** for the smaller weight. Boundary B contains the free body diagram for the larger weight.

We can see from free body diagram A that the out-of-balance forces acting in the direction of the acceleration are the tension T and the weight mg. In order to apply Newton's second law, we need to specify a direction for positive acceleration, so we will choose the direction of motion of the larger weight, as we did in the previous

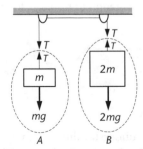

Figure 10.2   Creating a free body diagram by placing a cut in the system of Example 10.2.

example. The smaller weight is accelerating in the negative direction, so applying Newton's second law gives

$$F = ma$$

$$mg - T = -ma \qquad\qquad (10.7)$$

The tension and the acceleration are both unknown at this stage. If we now consider free body diagram B and carry out the same procedure we find that

$$F = ma$$

$$2mg - T = 2ma \qquad\qquad (10.8)$$

We now have two equations for the two unknowns, a and T. If we subtract Equation 10.7 from Equation 10.8, we get

$$(2mg - T) - (mg - T) = 2ma - (-ma)$$

$$mg = 3ma$$

$$a = \frac{g}{3} \ \mathrm{m\,s^{-2}}$$

This result is the same as we obtained in the previous example, which gives confidence in both calculations. However, we were asked to find the tension in the cable, which we could have found immediately if we had eliminated a from the equations instead of T. Now that we know a, we can simply substitute it in Equation 10.7, which then gives

$$mg - T = -m\left(\frac{g}{3}\right)$$

$$T = mg + m\left(\frac{g}{3}\right)$$

$$T = \frac{4}{3} \ mg\,\mathrm{N}$$

This too is a reasonable result because the tension is bigger than the smaller weight – so the out-of-balance force in free body diagram A must give an upward acceleration – and smaller than the bigger weight – so that the out-of-balance force in free body diagram B must give a downward acceleration.

Simple as it may be, this example illustrates the basic procedure required to solve a problem in dynamics. There are three steps:

➤ Define the problem and identify all the forces involved.
➤ Clarify the relationship between the various parts of the system to be analysed. This will involve drawing one or more free body diagrams, each of which should isolate a part from the system and show clearly all the forces acting on it.
➤ Calculate the required forces or accelerations using Newton's second law or equations derived from it. It is usually most convenient to apply Newton's second law in the same direction as the body will accelerate, if that can be identified in advance from the free body diagrams.

The first step identifies the forces involved in a problem. So far we have only considered forces arising from the mutual attraction between bodies, in particular, the gravitational pull of the earth. Next we consider other types of forces that may become important. Then we go on to see how Newton's second law can be developed into more convenient forms to accomplish the third step in a variety of applications.

Recap

■ **The mass of a body measures the amount of the material from which it is made. It is also a quantitative measure of its inertia, or its capacity to resist changes in its motion. It is not to be confused with the weight, which is the force exerted on the body by the gravitational field in which it exists.**

■ **For most engineering purposes the relationship between a body's motion and the forces acting on it are defined by Newton's second law.**

■ **SI units for mass, length and time are the kilogram, the metre and the second, respectively. The unit of force is defined with respect to Newton's second law and is known as the newton.**

■ **Problems in dynamics should be solved in a systematic manner through the use of free body diagrams and the application of Newton's second law or equations derived from it.**

➤ 10.3 More on forces

Forces are applied remotely or through direct mechanical contact. The force due to gravity is a remote force. All other types of force we shall consider will be contact forces. Some will have their origin in natural phenomena, such as gravitational forces, and some will be generated through the contrivance of man. Many manmade forces are created through the conversion of fuel into mechanical or electrical energy. Some forces precipitate motion, others resist it and, for obvious reasons, they are sometimes called **resistive forces.**

Resistive forces are often undesirable since many engineering applications are concerned with harnessing forces to produce motion. In some circumstances they are essential for the mechanism to produce the desired effect. Motor cars experience friction as a resistive force that is both undesirable and beneficial. It reduces the efficiency of the engine and is also a source of drag, but on the other hand it is essential for generating the tractive force produced by the interaction of the road and the wheels.

Forces tending to cause motion, which might be called **active forces**, are usually easy to identify and model because they tend to be gravitational forces, or forces applied by design. Normal reactions to these forces, as shown in Part I, are also straightforward. Active forces may squeeze a body, called **compressive forces**, or they may stretch it, called **tensile forces.** Some bodies, such as bars, can sustain both types of force. Others, such as ropes and cables, can only sustain tensile forces.

Resistive forces can be more difficult to deal with. It is not always obvious which of them are important and need to be considered. When they have been identified they can be difficult to model and sometimes it is not even possible to predict in which direction they are going to act. Resistive forces occur at interfaces between two solid bodies, between a solid body and a fluid, and between two fluids. We are mostly concerned with the first of these interfaces, where the resistive forces manifest themselves in the form of friction and the situation is relatively easy to model. We will also look briefly at the second interface, which is rather more complicated to model. The third is even more difficult to deal with and is beyond the scope of this book.

In any problem all of these forces have to be considered and included in a free body diagram, if they are relevant. In some problems, and this has been implicit in the ones we have considered so far, it is permissible to ignore some or all of the resistive forces; this is because they are so small compared to the others that their effects are negligible. In other problems they have to be included and may even dominate the system being considered. If they do have to be included, they have to be modelled, which means we have to be able to represent their behaviour in the form of equations leading to a quantitative description of their effect on the system. Before tackling more problems, we will spend some time considering some of these forces and how they may be modelled.

11 Friction

Friction is the tangential force developed between two surfaces when one of them slides, or is tending to slide, relative to the other. It is a resistive force that always acts in opposition to the direction of motion and invariably is responsible for energy losses within a system. Its magnitude depends on the nature of the surfaces involved. Friction manifests itself in several ways. The different types of friction include dry friction, fluid friction and internal friction.

Dry friction occurs when the surfaces of two solids in contact under a condition of sliding, or a tendency to slide, are unlubricated. This type of friction is also called **Coulomb friction** in recognition of the experimental work carried out by Coulomb in the eighteenth century, work which still forms the basis of contemporary models.

Fluid friction occurs when 'layers' of a fluid move relative to each other, where the word 'fluids' embraces both liquids and gases. It is convenient to think in terms of fluid layers while developing the concept of fluid friction. However, fluids are homogeneous materials in which the velocity changes continuously from one region of the fluid to another. It is therefore physically impossible for fluids to exist in distinct layers moving relative to one another with no transition region between them. It is also a law of nature that if a fluid is in contact with a solid surface, those particles of the fluid closest to the surface adhere to it and remain stationary. For this reason there is also a frictional force associated with the interface between a fluid and a solid, although the friction is actually generated within the fluid and should be thought of as fluid friction.

Internal friction occurs in all solid materials that are subjected to repeated loading and unloading. If they are made of highly elastic material, they recover from their deformation with very little energy loss due to internal friction. This is not the case for all materials, but internal friction can often be ignored and we will not consider it further.

11.1 Dry friction

Dry friction is one of the most commonly experienced forms of friction. Experiment has shown that the frictional force depends on the nature of the surfaces involved, although it is independent of the area of contact. There are aspects of friction which are still not totally understood, but the basic underlying physical mechanism involved can be illustrated by considering the behaviour of a block of mass m resting on a horizontal surface and subjected to a horizontal force T, as shown in Figure 11.1.

The force T is provided by the tension in a light cable connecting the block to a weight pan, where the pan and the weights have a total mass M. Because the system is

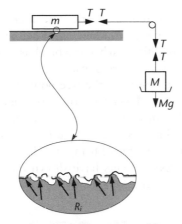

Figure 11.1 A block of mass m resting on a horizontal surface and subject to a force Mg. A magnified view of the mass resting on the horizontal surface is shown inset.

in equilibrium, the tension is equal to Mg and it can be varied by adding weights to the weight pan, so increasing the value of M.

The inset diagram represents a magnified view of a small area of contact between the block and the surface. At a microscopic scale the two surfaces are not flat but are composed of peaks and valleys. Contact between the two surfaces only occurs when the peaks of one surface touch the peaks or valleys of the other surface. The actual area of contact is therefore much smaller than the plan area of the block, which may be called the apparent area of contact. This explains why the frictional force is independent of the apparent area of contact.

The weight of the block acts as a distributed load over these contact points; and at each contact point there is a local reaction R_i acting at right angles to the local surface there. Each of the local reactions can be decomposed into a horizontal component F_i and a vertical component N_i. The frictional force F is the sum of the local horizontal forces, $F = \sum F_i$ and the total normal reaction of the sum of the local vertical forces $N = \sum N_i$.

The forces acting on the block can therefore be represented by its weight mg and the normal reaction N, opposing one another in the vertical plane, and the applied force T and the frictional force F, acting in opposition to one another in the plane tangential to the two surfaces. The free body diagram for the block is shown in Figure 11.2.

When the applied load is zero the block attains an equilibrium position in which the local horizontal forces sum to zero and the local vertical forces exactly balance the weight of the block. If the force T is gradually increased by the incremental addition of

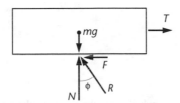

Figure 11.2 The free body diagram for the block.

weights to the weight pan, the block remains static until a critical point is reached when the addition of one more weight sets it in motion. When the block is stationary, friction is said to act under **static conditions**; when the block is moving, friction is said to act under **kinetic conditions**, for obvious reasons.

Under static conditions, as we can see from the free body diagram in Figure 11.2, the applied force T must always be balanced by the frictional force $F = F_s$, so F_s increases linearly with T, where the subscript s designates static conditions. A typical curve for the frictional force plotted against the applied force is shown in Figure 11.3 in which F_{smax} is the maximum value capable of being sustained by the frictional force before the block begins to move. Experiments have shown that, for a given pair of surfaces, F_{smax} is proportional to the normal reaction N, so

$$F_{max} \propto N$$

$$F_{max} = \mu_s N \tag{11.1}$$

The coefficient of proportionality, μ_s, is known as the **coefficient of static friction** and it can easily be determined experimentally. Note that Equation 11.1 is only valid when motion is pending and the coefficient of static friction can only be used to calculate the frictional force under these specific conditions.

When the block begins to move, then **kinetic conditions** apply. Under kinetic conditions the local contacts occur more nearly along the tops of the peaks, so the local reactions are more nearly vertical and their horizontal components are comparatively smaller than under static conditions. Consequently, once the block is moving, the force opposing the motion, F_k, is usually somewhat smaller than F_{smax}, the force required to induce the motion in the first place.

Under static conditions the frictional force is equal and opposite to the applied force $T = Mg$; this is because the system is in equilibrium. Experiments have shown that the frictional force under kinetic conditions is also proportional to the normal reaction, so

$$F_k = \mu_k N \tag{11.2}$$

where μ_k is known as the **coefficient of kinetic friction**.

The motion induced in the block is dictated by the out-of-balance horizontal force $(T - F_k)$. Adding weights to the weight pan increases tension which, in turn, increases the out-of-balance force and leads to larger accelerations. The curve shown in Figure 11.3 is typical of the results that may be obtained if a series of experiments were

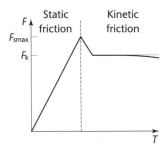

Figure 11.3 Frictional force plotted against applied force: the graph divides into two regions, static and kinetic.

undertaken in which the weights in the weight pan are successively increased up to and beyond the point where kinetic conditions occur.

The frictional force F_k appears to remain more or less constant for low values of T then gradually decreases as T increases. This reflects the fact that, as the velocity of the block increases, the kinetic frictional coefficient μ_k decreases slightly, and when high velocities are reached the effect of lubrication by an intervening fluid film may become appreciable. Unlike coefficients of static friction, values of μ_k are difficult to determine and assign with a high degree of certainty because not only do they depend on the velocity but also on the exact condition of the surfaces involved.

Figure 11.2 shows that the direction of the overall resultant force R can be defined with reference to the overall frictional and normal forces, F and N respectively. The angle between the resultant force and the direction normal to the surfaces, ϕ, is defined by the relationship

$$\tan \phi = \frac{F}{N} \tag{11.3}$$

where ϕ is known as the angle of friction.

When the frictional force reaches its limiting value, and motion is about to occur, ϕ reaches its maximum value, ϕ_s. If there is relative motion between the surfaces involved, the frictional force remains at a more or less constant value, F_k. Under these conditions the following equations hold

$$\tan \phi_s = \frac{F_{smax}}{N} = \mu_s$$

$$\tan \phi_k = \frac{F_k}{N} = \mu_k \tag{11.4}$$

where the **angle of static friction**, ϕ_s, and the **angle of kinetic friction**, ϕ_k, can be found by taking the appropriate arctangents. If there is no relative motion between the surfaces, the angle of friction is less than or equal to ϕ_s; if there is, the angle of friction is equal to ϕ_k.

Example 11.1

It is proposed to haul a 100 kg block up a 30° incline using the pulley system and 50 kg weight shown in Figure 11.4. Will the block actually move up the slope if the coefficient of static friction μ_s is 0.2?

Solution   Suppose initially that the block is in equilibrium but just about to move. From components in a direction perpendicular to the plane, we have

$$N - mg \cos 30° = 0$$

which gives a reaction N of 849.57 N.

The maximum frictional force that can develop is given by Equation 11.1 as

$$F_{smax} = \mu_s N = 0.2 \times 849.57 = 169.914 \text{ N}$$

Components parallel to the plane give

$$2T - F_{smax} - mg \sin 30° = 0$$

The tension in the cable is therefore

$$T = \frac{1}{2}(169.914 + 100 \times 9.81 \times \sin 30°) = 330.21 \text{ N}$$

The free body diagram for the weight shows that the force the weight experiences due to gravity is equal to g times its mass, which is $50 \times 9.81 = 490.5$ N. Since this is greater than the tension in the cable that has just been calculated, assuming the block is on the point of moving, the block must actually move up the slope.

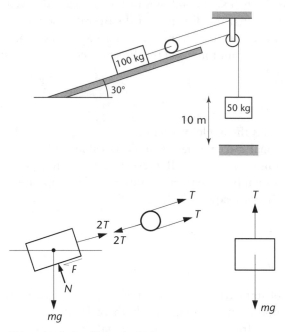

Figure 11.4   See Example 11.1.

➤   11.2   Rolling resistance

If a circular cylinder rolls along a flat surface, it experiences no resistive force unless there occurs slipping or distortion of the cylinder or the surface. In practice the surface is often distorted. This situation is illustrated in Figure 11.5, which also shows the

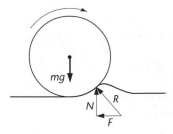

Figure 11.5   Rolling resistance: distortion of the surface gives the reaction a horizontal component.

forces involved. Due to the distortion of the surface, the reaction has a horizontal component which manifests itself as an apparent frictional force.

Like the true frictional force, the rolling resistance force is proportional to the normal load. If a coefficient of rolling resistance is defined in the same way as for sliding friction, its value tends to be relatively small. For this reason mechanisms are often designed so as to replace sliding friction with rolling friction.

➤ **Example 11.2**

Example 11.1 required a tension of 330.21 N to be developed in the cable for the block just to move. Suppose the block were now loaded on to a light wheeled trolley of mass 20 kg, and suppose the trolley were then attached to the same pulley system. What tension would be required in the cable for the trolley just to move if the rolling resistance were 1% of the normal load?

Solution   Following the same steps as in the previous example, and considering components in the direction normal to the slope, we have

$$R - (m_{block} + m_{trolley})g \cos 30° = 0$$

which gives a reaction of 1019.49 N. The rolling resistance F_{RR} is 1% of the reaction, which is 10.19 N. Resolving parallel to the slope, as before, leads to

$$2T - F_{RR} - mg \sin 30° = 0$$

which gives the tension in the cable as

$$T = \frac{1}{2}(10.19 + 120 \times 9.81 \sin 30°) = 299.40 \text{ N}$$

Even though the trolley increased the mass of the load to be hauled up the slope by 20%, the force required just to move it (299.40 N instead of 330.21 N) was reduced by nearly 10% by replacing sliding friction with rolling resistance.

11.3   Fluid friction and fluid drag

Fluid friction is a difficult phenomenon to understand and to model theoretically. Fluid mechanics is a field in its own right, embracing the phenomenon of fluid friction, which will only be discussed briefly here. Fluid friction is a resistive force of great importance in many engineering applications. Here it will be considered in the context of the **drag** experienced by a body moving in relation to a fluid in which it is immersed. That is, either the body or the fluid or both could be in motion. The fluid in question will be either air or water, and the motion will be relatively low speed.

If water flows slowly along a wide straight horizontal channel of constant cross-section, it is possible (though not strictly correct) to visualize it as being composed of many thin layers parallel to one another and the direction of flow. Along the centreline of such a channel, as a matter of observation, the further the fluid is from the bottom, the faster it moves. This is because the molecules of the layer adjacent to the bottom adhere to it and are stationary. Between the bottommost layer and the next, and between all adjacent layers, there is a fluid frictional force operating so as to oppose their relative motion. The lower layer tends to slow down the upper layer, and vice versa, producing a shearing force acting in a direction tangential to their interface. Newton proposed that the magnitude of the shearing force f_τ per unit area ΔA of the layer over which it acts, known as the **shear stress** τ is proportional to the rate at which the velocity U of the flow changes with distance y from the bottom, which for the simple flow we are considering can be represented as dU/dy.

$$\tau = \frac{f_\tau}{\Delta A} = \mu \frac{dU}{dy} \qquad (11.5)$$

where the coefficient of proportionality, μ, known as the coefficient of viscosity, at any given temperature is a characteristic of the fluid and constant throughout it.

In this model of the fluid behaviour, the fluid frictional forces depend on the viscosity of the fluid and the local rate of shear, characterized by the velocity gradient, $\partial U / \partial y$. Large velocity gradients correspond to high rates of shear and large frictional forces. At the interface of a fluid and a solid, the local frictional force depends on the velocity gradient as $y \rightarrow 0$. The overall frictional drag experienced by a body moving relative to the fluid is the sum of the local components of the frictional force opposing the motion over the surface of the whole body. A fluid whose real behaviour is well described by this model, such as water, is known as a **Newtonian** fluid.

The model becomes more complicated as the flow itself becomes more complicated. It has been described in terms of a simple straight and parallel one-dimensional flow. Extra components of the shear stress have to be introduced to model a fully three-dimensional flow. In the nineteenth century Reynolds demonstrated that the nature of fluid flow is yet more complicated than was assumed by Newton even for three-dimensional flows. The flow of fluids can be divided into two regimes: a **laminar** regime in which the flow remains well ordered and which is well described by Newton's model, and a **turbulent** regime in which the coefficient of proportionality, μ, varies throughout the flow field. The fluid particles in laminar flow in a straight pipe follow straight paths as they pass through it. If the velocity of the flow is increased, a point is reached at which the flow becomes chaotic and the fluid particles no longer travel in straight lines but have random transverse fluctuations superposed on their mean forward motion.

Although Newton's basic model has to be modified to capture this type of behaviour, as a rough guide, larger velocity gradients can still be associated with larger frictional forces. The mixing of the fluid is facilitated by turbulence, so the velocities increase more quickly with distance from the boundary than for laminar flows. This in turn leads to greater velocity gradients. Consequently, larger frictional forces are associated with turbulent flows than with laminar flows.

The modifications to Newton's model required to account for turbulence complicate it still further, and fully developed three-dimensional turbulent flows are very difficult to describe or to solve theoretically. In many engineering applications, theoretical solutions are not even possible and designers resort to the aid of experiments.

The fluid frictional forces on a body act in a direction opposing its motion and tangential to its surface. A body fully or partially submerged in a fluid also experiences forces due to the pressure of the fluid acting on it in a direction normal to its surface. Similarly, if the body and the fluid are at rest, the pressure is distributed evenly over the body in a horizontal direction and increases vertically with depth of fluid. The variation in the vertical pressure is responsible for a net upward force called the buoyancy force. If the body is moving relative to the fluid in the horizontal direction, the horizontal pressure distribution can become modified so there is a net force opposing the motion.

The frictional force opposing the motion is sometimes called **skin friction drag**, the pressure force **form drag** and their combination **profile drag.** The relative importance of the different types of drag depends upon the nature of the body itself. Bluff bodies, those that present a large projected area to the flow, such as circular cylinders, interact with the flow so as to produce large differentials in pressure over their surface. In these cases the profile drag consists largely of form drag. Streamlined bodies, on the other hand, cause comparatively little disturbance to the flow and their profile drag tends to be composed mostly of skin friction drag.

According to dimensional analysis and confirmed by experiment, over the range of interest for most engineering applications, the drag on a body moving through a fluid depends on the nature of the fluid, the nature of the body (respectively characterized by density ρ and volume) and the square of the body's velocity, V. For reasons that will not be discussed here, it appears to be convenient to express this mathematically in the form

$$F_D = C_D \left(\frac{1}{2} \rho V^2 \right) A \tag{11.6}$$

where A is the projected area of the body in the direction of the flow. For a circular cylinder in a transverse flow $A = LD$, where L and D are its length and diameter.

The coefficient of proportionality, C_D, is known as the **drag coefficient** and is written as

$$C_D = \frac{F_D}{\frac{1}{2} \rho V^2 AL} \tag{11.7}$$

Since the numerator and the denominator of the right-hand side both have dimensions of force, the drag coefficient is a pure number without dimensions. The important thing about a **non-dimensional coefficient** is that its value is the same for any consistent set of units. Thus measurements made using imperial units can be compared with measurements made using SI units by looking at a non-dimensional coefficient such as C_D. The value of the drag coefficient depends on the nature of the body, the nature of the fluid and the flow regime that exists. These attributes of the flow can be characterized by the non-dimensional grouping known as the **Reynolds number** $R_n = VD/v$, where D and V represent the scale of the body (e.g. its diameter) and the fluid velocity respectively and v is the kinematic viscosity, μ/ρ. The drag coefficient for a smooth sphere in a steady uniform and unidirectional current at a given Reynolds number is fixed, no matter what combination of D, V and v makes up that Reynolds number.

Many experiments have been carried out on spheres, and many other bodies of engineering interest, to determine the drag coefficient over a wide range of Reynolds numbers. The results of these experiments have been recorded and are widely available in textbooks and scientific publications, mostly as curves of drag coefficient plotted against Reynolds number. The drag coefficient for any body in a given fluid flow can be found from one of these curves and the drag force calculated from Equation 11.6.

In the case of dry friction, if two solid surfaces are in contact, the force required to set one in motion relative to the other is the force required to overcome the local reactive forces of the roughness projections of the surfaces involved and is independent of the apparent contact area. If the two surfaces are separated by sufficient fluid to prevent contact between them, the force required to set one of them in motion is the force required to overcome the shear forces in the fluid, which are proportional to the apparent area of contact and which increase with speed. Behaviour of this sort is sometimes called **viscous friction** for obvious reasons.

➤ **Example 11.3**

The coefficient of frictional resistance of a ship can be found from the equation

$$C_F = \frac{F}{\frac{1}{2}\rho V^2 S} = \frac{0.075}{(\log_{10} R_n - 2)^2}$$

Calculate the frictional force on a ship of length $L = 100\,\text{m}$ and wetted surface area $S = 3500\,\text{m}^2$ travelling at a speed $V = 10\,\text{m}\,\text{s}^{-1}$. The density ρ and kinematic viscosity v may be taken as $1000\,\text{kg}\,\text{m}^{-3}$ and $10^{-6}\,\text{m}^2\,\text{s}^{-1}$ respectively.

Solution   In order to calculate the frictional force it is necessary to know the frictional coefficient, which in turn requires knowledge of the Reynolds number:

$$R_n = \frac{VL}{v} = \frac{10 \times 100}{10^{-6}} = 10^9$$

The frictional coefficient is then given by

$$C_F = \frac{0.075}{(\log_{10} 10^9 - 2)^2} = 1.53 \times 10^{-3}$$

Finally, the frictional force can be calculated from

$$F = \frac{1}{2}\rho V^2 S C_F = \frac{1}{2} \times 1000 \times 10^2 \times 3500 \times 1.53 \times 10^{-3} = 267.75\,\text{kN}$$

The frictional force experienced by the ship travelling at $10\,\text{m}\,\text{s}^{-1}$ is 267.75 kN.

Recap

■ Dry friction, sometimes known as Coulomb friction, occurs when two unlubricated solid surfaces move, or tend to move, relative one to the other. The resulting force is tangential to their interface and acts so as to resist motion. The maximum value for static friction is proportional to the normal reaction. The frictional force between two surfaces in relative motion is also proportional to the normal reaction, although the coefficient of kinetic friction is less than the coefficient of static friction. The frictional force is independent of the apparent area of contact.

■ Rolling resistance occurs when a surface is distorted due to the action of a cylinder rolling over it in such a way as to cause resistance to the cylinder motion. Its magnitude is also proportional to the normal reaction. Mechanisms designed to replace sliding friction by rolling resistance generally lead to a reduction in resistive forces.

■ Fluid friction is a complicated phenomenon in which the frictional force depends on the area of contact and the magnitude of the relative motion of the surfaces involved. The drag on a body moving through a fluid is in part due to the frictional force existing at the fluid–body interface. These quantities are difficult to predict theoretically and are usually estimated from experimental data.

11.4 Problems

1. A block rests on an inclined plane. If motion is impending and the coefficient of static friction, μ_s, is 0.2, what is the angle of inclination of the plane?

2. If the coefficient of static friction is 0.2 for the arrangement shown in Figure 11.4, calculate the range over which the tension, T, can vary without the block moving.

3. If the block shown in Figure 11.4 is uncoupled from the pulley system what is the least horizontal force that must be applied to the block to prevent it from sliding down the slope?

4. A boat is being hauled up a slipway that is inclined at 20° to the horizontal using a winch whose cable is parallel to the slipway. If the boat has a mass of 2000 kg and μ_k is 0.4, what is the tension in the cable?

5. A uniform block of 100 kg mass having a square base of 1 m side rests on a horizontal surface with a coefficient of static friction of 0.25. If it is wished to move the block by sliding it without toppling it

by the application of a horizontal force, what is the maximum height above the base that the force can be applied? (Hint: investigate toppling by taking moments about one edge of the base.)

6. A 15 tonne lorry develops a tractive effort of 16.5 kN as it travels at a steady speed up a 1 in 10 incline. If the drag on the lorry due to air resistance is 300 N, ignoring all other losses, calculate the rolling resistance of the lorry as a percentage of its normal reaction.

7. Calculate the air resistance (drag) of a car travelling at 30 m s^{-1} if it has a projected area of 1.9 m^2 and a drag coefficient, C_D, of 0.8. The density of air may be taken as 1.19 kg m^{-3}.

8. Calculate the frictional resistance of a 50th scale model of the ship described in Example 11.3 of this chapter if it is being towed in a resistance experiment at a velocity of 1.4 m s^{-1}. (Hint: use the same method but remember area scales as the square of the scale factor.)

12 Applying Newton's second law

Having introduced Newton's second law and some of the ideas associated with it, and having identified and discussed the nature of the various forces likely to be encountered in engineering applications, it is possible to use them to solve some problems. In the section on kinematics various types of motion were considered without reference to the forces that caused them. In dealing with Newton's second law, we are now in the realm of kinetics and interested in the relationship between motions and the forces that brought them about. As with the section on kinematics, we shall begin with rectilinear motion.

12.1 Rectilinear motion

Example 12.1

Find the tension in the hoisting cable of a lift weighing 700 kg if it is ascending with an acceleration of $1.5\,\mathrm{m\,s^{-2}}$ and contains an 80 kg passenger.

Solution The forces acting in this problem are the tension in the cable and the gravitational pull on the passenger and the lift. Since the lift is accelerating, the forces are out of balance. Newton's second law states that the rate of change of momentum of a body, which can be written as the product of its mass and acceleration when the mass remains constant, is equal to the sum of the forces acting on it. That is, taking the upward direction as being positive,

$$(m_{\text{lift}} + m_{\text{person}})a = T - m_{\text{lift}}g - m_{\text{person}}g \tag{12.1}$$

The tension T is the only unknown and can be found by rearranging the equation as

$$T = (m_{\text{lift}} + m_{\text{person}})a + (m_{\text{lift}} + m_{\text{person}})g$$

$$= (700 + 80) \times 1.5 + (700 + 80) \times 9.81 = 8822\,\mathrm{N} \tag{12.2}$$

The tension in the hoisting cable is therefore **8.822 kN**.

In this problem the body is not in equilibrium because the forces acting on it are out of balance. As we have seen, the sum of the out-of-balance forces is equal to ma. It is obviously possible to add another force, $f_{\mathrm{I}} = -ma$, so as to bring the system back to equilibrium. This force, which is equal and opposite to the resultant of the forces acting on the body, is known as the **inertia force**. Another approach to the problem, then, is to convert the unbalanced system to one in equilibrium by the addition of the inertia force and to treat it in accordance with the rules of statics. This approach is known as D'Alembert's principle.

In the present case, the inertia force $f_I = -(m_{lift} + m_{person})a$; and the tension, the weight of the person and the lift, and the inertia force are in equilibrium, so that

$$T - (m_{lift} + m_{person})g + f_I = 0$$

Substituting for the inertia force and rearranging the equation leads to

$$T = (m_{lift} + m_{person})g + (m_{lift} + m_{person})a$$

which is the same as Equation 12.2 and gives the same answer as before.

D'Alembert's principle will not be used again in this text but it is sometimes convenient to think in terms of an inertia force that has to be overcome in order to cause the acceleration of a body. And the idea of an inertia force crops up in other contexts. For example, the Reynolds number that was encountered in the previous section can be treated as the ratio of the inertia to the viscous forces acting in a fluid. Understanding this is of great assistance in understanding the behaviour of a viscous fluid.

> ## Example 12.2

As the lift in the previous problem neared the end of its journey, suppose the tension in the cable reduced to 8000 N. What would have been the reactive force of the floor beneath the passenger's feet?

Solution   In this problem it is convenient to consider first the forces acting on the lift and the passenger together, then the forces acting on the passenger alone. Applying Newton's second law gives

$$(m_{lift} + m_{person})a = T - (m_{lift} + m_{person})g$$

In this case the acceleration is the only unknown, so

$$a = \frac{T - (m_{lift} + m_{person})g}{(m_{lift} + m_{person})}$$

$$= \frac{8000 - (700 + 80) \times 9.81}{(700 + 80)} = 0.446 \text{ m s}^{-2}$$

The passenger moves with the lift and therefore has an acceleration of 0.446 m s^{-2}. Since the lift is ascending, the gravitational pull on the passenger is overcome by the reactive force of the floor on his or her feet propelling him upwards. Applying Newton's second law to the passenger alone gives

$$m_{person}a = F_{react} - m_{person}g$$

The reactive force of the floor is therefore given by

$$F_{react} = m_{person}a + m_{person}g$$

$$= 80 \times 0.446 + 80 \times 9.81 = 820 \text{ N}$$

The reactive force of the floor on the passenger's feet is **820 N**.

➤ **Example 12.3**

Example 11.1 showed that the weight and pulley system illustrated in Figure 11.4 would cause the block to move up the slope. If the weight were released from rest from a height of 10 m, what would be its velocity just before it hit the ground?

Solution It can be seen from Figure 11.4 that for every metre the block moves, the weight must move 2 m. Since these movements occur simultaneously, the weight must have twice the acceleration of the block. Assuming kinetic conditions and resolving considering components to the slope,

$$2T - \mu_k N - m_{block} g \sin 30° = m_{block} a \tag{12.3}$$

If we now consider the weight by itself, which we have already demonstrated is moving in a downward direction, we get

$$m_{weight} g - T = m_{weight}(2a)$$

where a is the acceleration of the block. Rearranging the equation gives

$$T = (g - 2a)m_{weight} \tag{12.4}$$

Substituting for T in the previous equation leads to

$$m_{block} a = 2(g - 2a)m_{weight} - \mu_k N - m_{block} g \sin 30°$$

Collecting terms in a on the left-hand side of the equation gives

$$(m_{block} + 4m_{weight})a = 2m_{weight} g - \mu N - m_{block} g \sin 30°$$

so the acceleration of the block is given by

$$a = \frac{1}{m_{block} + 4m_{weight}}(2m_{weight} g - \mu N - m_{block} g \sin 30°)$$

$$= \frac{1}{100 + 200}(2 \times 9.81 \times 50 - 0.17 \times 849.57 - 100 \times 9.81 \times \sin 30°)$$

$$= 1.153 \text{ m s}^{-2}$$

The acceleration of the weight a_{weight} is therefore $2a$, which equals 2.306 m s^{-2}. Since none of the forces acting on the weight change with time, its acceleration must be constant and the kinematic equations for constant acceleration can be used. In addition to the acceleration of the weight, its initial velocity, $v_1 = 0$, is known since it was released from rest. In order to find its velocity v_2 when it hits the ground an equation is required that connects v_2 with v_1, a_{weight} and the final displacement, s_2. From Section 7.6 it can be seen that the appropriate equation is given by

$$v_2^2 = v_1^2 + 2a_{weight}s$$

$$= 0 + 2 \times 2.306 \times 10 = 46.12 \text{ m}^2 \text{ s}^{-2}$$

On taking the square root of v_2^2 it can be seen that the weight hits the ground with a velocity of just under **6.8 m s^{-1}**.

12.2 Curvilinear motion

Example 12.4

A car of 1500 kg mass is travelling along a straight road at 50 kph. It passes through a dip in the road whose radius of curvature ρ is 50 m followed by a hump for which ρ is also equal to 50 m. Calculate the total reaction R on the wheels of the car as it passes through the bottom of the dip. How fast would the car be travelling if it were just about to lose contact with the road as it travelled over the top of the hump?

Solution   As the car passes the bottom of the dip it will experience a normal component of acceleration a_n directed towards the centre of curvature (upwards) and a tangential velocity v_t equal to its speed. Applying Newton's second law and resolving in a direction normal to the road gives

$$F = ma_n$$
$$R - 1500 \times 9.81 = 1500a_n \tag{12.5}$$

Since the tangential velocity and the radius of curvature are both known, the normal component of acceleration can be calculated from Equation 9.5.

$$a_n = \frac{v_t^2}{\rho} = \left(50 \times \frac{1000}{60 \times 60}\right)^2 \left(\frac{1}{50}\right) = 3.858 \text{ m s}^{-2}$$

Substituting for the normal component of acceleration in Equation 12.25 leads to

$$R = 1500 \times 9.81 + 1500 \times 3.858 = 20502 \text{ N}$$

The reaction on the wheels of the car is therefore **20.5 kN** as it passes through the dip in the road.

As the car passes the top of the hump in the road, it again experiences a normal component of acceleration directed towards the centre of curvature of the road, which in this case is downwards. Repeating the procedure of the earlier part of the question now gives

$$F = ma_n$$
$$mg - R = ma_n \tag{12.6}$$

If the car is travelling fast enough it will lose contact with the road surface as it passes the crest of the hump, and the reaction on the wheels will disappear. If R is zero, then as can be seen from Equation 12.6, the normal component of the acceleration must be equal to g. If a_n and ρ are both known, the tangential velocity can be calculated from Equation 9.5.

$$v_t = \sqrt{\rho a_n} = \sqrt{\rho g} = \sqrt{50 \times 9.81} = 22.15 \text{ m s}^{-1}$$

Therefore the car will lose contact with the surface of the road if it travels faster than **80 kph** as it passes the brow of the hill.

➤ **12.3 Periodic motion**

➤ **Example 12.5**

A body whose mass is 25 kg is constrained to run along straight horizontal frictionless guide rails. It is connected to a fixed point by a light horizontal spring whose stiffness is 20 kN m^{-1}. On being released after having been given an initial displacement, the body oscillates to and fro horizontally with simple harmonic motion. Calculate the period of the motion.

Solution The only horizontal component of force acting on the body is exerted by the spring. It always acts in opposition to the direction of the motion and, as we will see in Section 13.3, its magnitude is linearly proportional to its displacement s. The force acting on the body due to the spring can therefore be written as

$$F = -ks \tag{12.7}$$

Applying Newton's second law in the horizontal plane and combining it with Equation 12.7 leads to

$$F = ma$$
$$-ks = m\ddot{s} \tag{12.8}$$

The displacement and acceleration for a body undergoing simple harmonic motion are given by Equations 9.23 and 9.25. Combining them with Equation 12.8 gives

$$-k\hat{s} \sin \omega t = m(-\omega^2 \hat{s} \sin \omega t) \tag{12.9}$$

where ω is the circular frequency of Equation 9.15. The period of the motion is T, where

$$T = \frac{2\pi}{\omega} \tag{12.10}$$

Rearranging Equation 12.9 leads to

$$\omega^2 = \frac{k}{m}$$

Therefore the period is

$$T = 2\pi\sqrt{\frac{m}{k}} = 2\pi\sqrt{\frac{25}{18\,000}} = 0.234\,\text{s}$$

The body oscillates backwards and forwards with a period of **0.234 s**.

The problems considered so far have all involved the direct application of Newton's second law. For some problems it is often more convenient to apply the law indirectly using concepts and equations derived from it. These approaches will be described in the following chapter.

Recap

■ The inertia force is equal and opposite to the resultant of the forces acting on the body and can be thought of as the force that must be overcome in order to bring about an acceleration of the body. D'Alembert's principle converts a problem in dynamics to a problem in statics by the addition of an inertia force to the system under consideration.

■ In problems involving rectilinear motion, it is often convenient to consider components parallel to, and at right angles to, the direction of motion and to apply Newton's second law in those directions. In one of these directions there will be no out-of-balance force.

■ In problems involving curvilinear motion there are two components of acceleration: one in line with the direction of motion (as in rectilinear motion) and the other directed at right angles to it and towards the centre of curvature. It is possible to have unbalanced forces in both directions.

■ In simple harmonic motion the out-of-balance force always acts in opposition to the direction of the motion.

12.4 Problems

1. A mass of 1 kg is suspended from a spring balance in a lift. What is the reading on the spring balance (a) when the lift is stationary, (b) when it is ascending with an acceleration of $1.19\,\mathrm{m\,s^{-2}}$, (c) when it is ascending with a deceleration of $1.19\,\mathrm{m\,s^{-2}}$, and (d) when it is descending with a deceleration of $1.19\,\mathrm{m\,s^{-2}}$? Take the acceleration due to gravity to be $9.81\,\mathrm{m\,s^{-2}}$.

2. An 8 Mg lorry is designed to tow a 16 Mg trailer. If it develops a tractive force of 15 kN as it pulls away from rest, calculate the tension in the tow bar and the acceleration of the unit.

3. A 60 tonne locomotive is pulling some carriages, of mass 240 tonnes, down a gradual incline of 1 in 400 when its driver applies the brakes. How far does the train travel before the speed is reduced from 80 kph to 50 kph if the track resistance is 50 N for every tonne mass of the train and the braking force is 60 kN?

4. A particle of 2 kg mass is moving horizontally with a velocity of $5\,\mathrm{m\,s^{-1}}$ when a horizontal force, F, given by $F = 5 + t^2$ is applied in the direction of motion. What will be the velocity of the particle when the time t is 2 s?

5. A car, of mass 1000 kg, on a fairground roundabout is connected to a central spigot by a bar of 4 m length. What is the tension in the bar when the car is taking 2 s to make one complete horizontal circuit?

6. A particle travels with constrained motion along a vertical guide shaped like an inverted U. If the mass of the particle is 0.5 kg, what is the force exerted on the particle by the guide as it passes the highest point with a horizontal velocity of $4\,\mathrm{m\,s^{-1}}$ where the radius of curvature is 0.5 m?

7. A car rounding a bend of 120 m radius of curvature at 90 kph is just on the point of sliding. What is the coefficient of friction between the tyres and the road?

8. A car travels a distance of 200 m as it rounds a left handed bend of 400 m radius of curvature immediately followed by a right handed bend of 100 m radius of curvature. During this manoeuvre it slows down at a uniform rate from an initial speed of 90 kph to a final speed of 50 kph. If the mass of the car is 1500 kg, calculate the total horizontal force exerted by the road on the car tyres (a) on the left handed bend, (b) on the right handed bend and (c) at the point of inflection where the two bends meet.

9. A body of 20 kg mass is undergoing simple harmonic motion in a straight line, its greatest excursion from its mid position being 0.75 m. If the body completes 40 cycles in one minute, calculate the force acting on the body (a) at the beginning of its travel, and (b) three eighths of the way through its cycle.

10. A wave rider buoy of 975 kg mass is attached to the sea bed by a cable that is taught but unstretched in calm water. If the more complex interaction between the buoy and the sea can be ignored, and it can be assumed that the buoy follows exactly the profile of the waves, calculate the tension in the cable when the buoy is at the peak of a wave of amplitude 5 m and period 15 s if it is subject to a buoyancy force of 20 kN.

13 Work and energy

In the kinetics problems considered so far, information regarding the kinematics of bodies has been found from a knowledge of the out-of-balance forces acting on the system, or information concerning the forces has been obtained from a knowledge of the motions involved. Both cases have involved the determination of the acceleration and the application of Newton's second law at some instant. If the cumulative effect of the out-of-balance forces over a period of time was of interest, this was taken care of by integrating the acceleration in the equations describing the kinematics of the system.

When dealing with problems in which forces are applied over a distance, or over an interval of time, it is possible to integrate Newton's second law directly, so the acceleration never has to be calculated explicitly. This approach leads to two classes of problems. The first, involving integration with respect to distance, leads to the concepts of **work** and **energy**. The second, involving integration with respect to time, leads to the concepts of **impulse** and **momentum**. Although the introduction of these ideas may seem to complicate an already demanding subject still further, they often lead to simple and elegant solutions of otherwise difficult problems.

The concepts of work and energy at least are not entirely alien, and they will be dealt with first. If a force is applied to a body, or a system, and motion occurs in the direction of the force, work is done. Work is achieved through the expenditure of energy. If a body, or system, has energy stored within it, it is capable of doing work. Energy can take many forms and can be classified under a variety of headings such as chemical, thermal, solar, nuclear, mechanical and electrical. An important aspect of engineering is the transformation of energy from one form to another form that can be harnessed to do useful work. For example, the chemical energy bound up in petrol can be transformed into the mechanical energy providing the tractive force for a motor vehicle.

The following sections will be devoted to mechanical energy. Mechanical energy is required to bring about the motion to achieve mechanical work. Similarly, a body already in motion is a source of energy that is available for transformation into mechanical work. For example, the energy in moving air can be harnessed to drive a wind turbine. The energy of a body arising from its motion is known as **kinetic energy**. In certain circumstances mechanical energy can be stored in a body by doing work on it. An example of this is when work is done on a body by raising it against gravity up to a certain height. The energy required for it to move back to its original position, which can be thought of as being stored in the body, can be released simply by dropping it. Another example is found in applying a force to extend an elastic spring through a certain distance. On releasing the spring, if none of the stored energy is used

153

for other purposes such as generating heat, it will return to its original position. A body that can be thought of as having stored energy has the potential for doing work. The energy is known as **potential energy** and its magnitude depends on its position relative to some given datum. The concepts of work and energy will be explored further in the following sections.

➤ 13.1 Work, power and efficiency

Work is done when a body moves some distance in response to the application of a force, or system of forces. The larger the force, or the greater the distance, the greater the work done. Work may be defined as the product of the displacement of a body and the component of the resultant force acting on it in the direction of the displacement.

If a particle moves along a curved path, as shown in Figure 13.1, under the action of a variable force F, the work done ΔU during a displacement Δs is given by

$$\Delta U = F \cos \theta \, \Delta s$$

The total work done on the particle as it moves from O to P is therefore

$$U = \int_0^P F \cos \theta \, \mathrm{d}s \tag{13.1}$$

Work is a scalar quantity, but if the applied force opposes the motion (i.e. $\pi/2 < \theta < 3\pi/2$) it takes a negative sign. Similarly, if the applied force acts at right angles to the motion $\cos \theta = 0$ and the work done is zero. The SI units of work are newton-metres (N m), more commonly known as **joules** (J).

➤ Example 13.1

In Example 12.3 the component of the resultant force acting parallel to the slope, F, is given by the left-hand side of Equation 12.3:

$$F = 2T - \mu_k N - m_{block} g \sin 30° = m_{block} a$$

The direction of F is also the direction of the motion, so the $\cos \theta$ term of Equation 13.1 becomes $\cos 0 = 1$. Since all the variables in the problem remain constant in time, the force

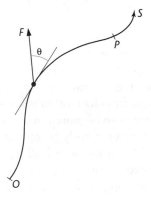

Figure 13.1 A particle moving on a curved path under the influence of a variable force.

applied to the block in the direction of its motion also remains constant, and the work done on it as it moves a distance s is simply given by $U = F \times s$. As was seen in the example, if the weight falls 10 m the block moves a distance s of 5 m up the slope. And Equation 12.4 shows the tension T is 375.2 N. The work done on the block as the weight falls from a height of 10 m to the ground is therefore given by

$$U = \int_0^s F \cos \theta \, ds = \int_0^s F \times 1 \, ds = F \int_0^s ds = Fs$$

$$= (2 \times 375.2 - 0.17 \times 849.57 - 100 \times 9.81 \times \sin 30°) \times 5$$

$$= 115.4 \times 5 = 577 \, J$$

Machines are designed to do work. In assessing the performance of machines it is necessary to have an objective measure of how well they carry out their function according to some given criteria. One criterion of great importance is how quickly a machine does work. The time rate of doing work, P, is known as **power**. A more powerful machine does more work in a given period of time, or the same amount of work in a shorter time, than a less powerful machine. The unit of power is the **watt** (W) and it is defined as 1 joule per second.

➤ **Example 13.2**

On average, how much power did the weight and pulley system in the previous example develop in pulling the block up the plane?

Solution   The work done on the block as it travelled a distance of 5 m with a constant acceleration of 1.153 m s^{-2} starting from rest ($v_1 = 0$) was 577 J. From Section 8.5 on the kinematics of rectilinear motion we know that

$$s = v_1 t + \frac{1}{2} at^2$$

The work on the block was done over a period t given by

$$t = \sqrt{\frac{2s}{a}} = \sqrt{\frac{2 \times 5}{1.153}} = 2.95 \, s$$

The average power developed by the weight and pulley system is the work done divided by the time taken to do it, which is equal to 195.59 W.

There is another approach that can be taken to calculating power when a constant force is involved.

$$P = \frac{dU}{dt}$$

$$= \frac{d}{dt}(Fs) = F\frac{ds}{dt} = Fv \tag{13.2}$$

In this problem the applied force F is 115.4 N and the average velocity is the distance divided by the time taken to carry out the work, 5/2.95 m s^{-1}. The average power developed is therefore given by

$$P = Fv = 115.4 \times \frac{5}{2.95} = \mathbf{195.6 \, W}$$

As might be expected, this is identical to the earlier answer.

➤ **Example 13.3**

A ship 100 m long and of wetted surface area 3500 m^2 is travelling at a speed of 5 m s^{-1}. Calculate the power its engine has to deliver in order to overcome the frictional resistance of the water.

Solution In Example 11.3 on fluid friction it was found that the frictional resistance of a ship of the same dimensions travelling at 5 m s^{-1} was 267.75 kN. Substituting these figures into Equation 13.2 gives

$$P = Fv = 267.75 \times 1000 \times 5 = 1.338 \times 10^6 \, \text{W}$$

The engine has to develop a power of 1.3 MW in order to overcome the frictional resistance of the water on the ship when it travels at 5 m s^{-1}.

There are various criteria upon which the performance of a machine can be assessed. A shipowner is interested in profit and in most cases this is increased if his or her ship travels at high speed. However, the shipowner may be content with a less powerful engine and lower speed if it is cheaper to run. One of the factors that determines the running cost of an engine is its **efficiency**. Efficiency is a measure of how much of the energy that is available as input is converted into useful work.

Mechanical efficiency, η_m can be defined as the ratio of the useful work done to the actual work input in a given time. The input work and the output work are carried out in the same time, so the efficiency can be written as

$$\eta = \frac{\text{output power}}{\text{input power}} \tag{13.3}$$

In all machines work, and therefore power, is lost mainly through friction and the generation of heat. In addition there may be electrical and thermal losses leading to the concepts of electrical efficiency, η_e and thermal efficiency, η_t. The overall efficiency in such instances is given by

$$\eta = \eta_m \eta_e \eta_t \tag{13.4}$$

Similarly the individual components of a machine or system have their own individual efficiencies and the overall efficiency is obtained by multiplying them together.

➤ **13.2 Kinetic energy**

Work is done when a body has moved a certain distance under the application of a force. The relationship between the force and the motion at any given instant is given by Newton's second law. If the mass of the body remains constant

$$F = m\ddot{x}$$

As seen in Section 8.5, acceleration can be expressed in terms of velocity and displacement, so Newton's second law can be rewritten as

$$F = mv\frac{dv}{ds}$$

The cumulative effect of the force over the whole displacement of the body, rather than at any one instant, can be obtained by integrating the equation with respect to the displacement:

$$\int_0^s F \, ds = \int_{v_1}^{v_2} mv \, dv$$

$$= \frac{1}{2}mv_2^2 - \frac{1}{2}mv_1^2$$

The left-hand side of the equation is clearly the work done U, so

$$U = \frac{1}{2}mv_2^2 - \frac{1}{2}mv_1^2 \tag{13.5}$$

The total work T that must be done on a particle to bring it to a velocity v starting from a state of rest ($v_1 = 0$) is given by

$$T = \frac{1}{2}mv^2 \tag{13.6}$$

This quantity T is known as the kinetic energy and represents the energy a particle has by virtue of its motion. Kinetic energy is always a positive scalar quantity and it has the same units as work, N m or joules.

The right-hand side of Equation 13.5 represents the change in kinetic energy ΔT while the work is being done. The equation can therefore be written

$$U = \Delta T \tag{13.7}$$

This equation is known as the **work-energy equation**; it states that the total work done by all the forces acting on a particle during an interval of its motion equals the corresponding change in kinetic energy of the particle.

➤ **Example 13.4**

Calculate the work done when a 20 tonne railway wagon travelling at $4 \, \text{m s}^{-1}$ is brought to rest by a buffer.

Solution   The work done is equal to the change in kinetic energy. The final velocity v_2 is zero, therefore

$$U = \Delta T = \frac{1}{2}mv_2^2 - \frac{1}{2}mv_1^2 = \frac{1}{2}m(v_2^2 - v_1^2)$$

$$= \frac{1}{2} \times 20 \times 10^3 (0 - 4^2) = -160\,000 \, \text{J}$$

Therefore, the work done in bringing the wagon to rest is **160 kJ** against the direction of motion, hence the negative sign.

This problem could have been solved using one of the equations of motion, Newton's second law and the equation for work, Equation 13.1. It would have involved satisfying the equations simultaneously for the unknown force, the distance over which it was applied and the deceleration it brought about. This is all taken care of automatically by the work–energy equation, which is a much simpler method of solution, not to mention more elegant.

Before continuing to demonstrate the advantages of solving problems using the work–energy equation, it is useful to consider the other form of mechanical energy, potential energy.

➤ ## 13.3 Potential energy

Sometimes the work done on a body in moving it from one location to another can be thought of as being stored in the body. The stored energy available for further work, known as potential energy, depends upon its position. The concept of potential energy can be applied in several different contexts, but first consider the potential energy of a spring being stretched or compressed through the application of a load F. If the spring is light and elastic it often exhibits the linear behaviour shown in Figure 13.2. Suppose the load F produces a deflection δ. The load required to give the spring a unit deflection, given by the slope of the graph, is known as the **spring stiffness**, K and

$$K = \frac{F}{\delta} \tag{13.8}$$

The force required to produce any deflection x is Kx, so the work done to achieve a displacement of δ is given by

$$\int_0^\delta F \, \mathrm{d}x = \int_0^\delta Kx \, \mathrm{d}x$$

$$= \left[\frac{1}{2}Kx^2 \right]_0^\delta = \frac{1}{2}K\delta^2 \tag{13.9}$$

The work done in extending or compressing the spring through a displacement of δ can be considered as stored by the spring and, if there are no losses in the system due to effects such as friction, it can be recovered when the spring is released.

If the potential energy of the spring, V_e (where the subscript denotes elastic) is defined as the recoverable energy stored by it when work is done on it, then

$$V_e = \frac{1}{2}K\delta^2 \tag{13.10}$$

Suppose a vertical light elastic spring fixed at its upper end experiences a deflection δ_1 due to a weight pan attached to its lower end. If the extension is increased to δ_2 by the

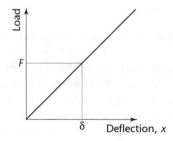

Figure 13.2 Linear behaviour of a light elastic spring.

application of a force F caused by placing a weight in the pan, the **work done on the spring by the weight** is given by

$$\int_{\delta_1}^{\delta_2} F \, dx = \int_{\delta_1}^{\delta_2} Kx \, dx$$

$$= \frac{1}{2}K\delta_2^2 - \frac{1}{2}K\delta_1^2 = \Delta V_e$$

The work done on the spring as its extension changes from δ_1 to δ_2 is equal to the change in its potential energy. If the weight exerts a force F on the spring, the spring exerts an equal and opposite force $-F$ on the weight. The **work done on the weight by the spring** is therefore equal to the negative change in potential energy

$$U_e = -\Delta V_e \tag{13.11}$$

This equation is the work–energy equation considering only the work done by the spring and the spring potential energy.

▶ **Example 13.5**

The work done by a buffer (a spring-mounted stop) in bringing a railway wagon to rest has been calculated to be $-160\,\text{kJ}$. Given a buffer spring stiffness of $8 \times 10^5\,\text{N m}^{-1}$ and assuming that all the work is done by the buffer, calculate its deflection.

Solution   The problem is simply solved using Equations 13.10 and 13.11.

$$U_e = -\Delta V_e = -\left(\frac{1}{2}K\delta_2^2 - \frac{1}{2}K\delta_1^2\right)$$

Before the wagon starts to compress the spring, its deflection δ_1 is zero, therefore

$$-160 \times 10^3 = -\left(\frac{1}{2} \times 8 \times 10^5\delta_2^2 \quad 0\right)$$

and the deflection of the spring is given by

$$\delta_2 = \sqrt{\frac{160 \times 10^3}{\frac{1}{2} \times 8 \times 10^5}} = \textbf{0.632 m}$$

A similar expression to Equation 13.11 can be developed for the work done by the gravitational force field.

▶ ## 13.4 Gravitational potential energy

The gravitational potential energy V_g of a particle is defined as the work done on the particle against the gravitational field in raising it to a height h above some reference plane where the gravitational potential energy is taken as zero, that is

$$V_g = mgh \tag{13.12}$$

If a particle is raised from a point where $h = h_1$ to a point where $h = h_2$ the work done by the gravitational field on the particle is given by

$$U_g = -mg(h_2 - h_1)$$

because the gravitational field is acting in the opposite direction to the motion of the particle.

The change in the gravitational potential energy is given by

$$\Delta V_g = V_g(\text{when } h = h_2) - V_g(\text{when } h = h_1)$$

$$= mgh_2 - mgh_1 = mg(h_2 - h_1)$$

The work done on the particle by the gravitational field is therefore equal to the negative change in its gravitational potential,

$$U_g = -\Delta V_g \qquad\qquad (13.13)$$

Forces such as the elastic and gravitational forces described in this section are known as **conservative forces** because the work done against them is conserved it is fully recoverable. The work done against a conservative force depends only on the net change of position, not on the particular path followed in reaching the new position. If a stone were dropped from the top of the tower of Pisa, its change in gravitational potential energy would be just the same when it reached the bottom as it would have been if it had been carried down by the staircase. More often than not, forces are not conservative. For example, frictional forces, whether associated with dry friction or fluid drag, do not have these attributes.

➤ 13.5   Conservation of energy

The work–energy equation can be modified to include potential energy. Suppose U is the work done by all the forces acting on a body other than conservative forces, which do work U_c. Since the total work done is equal to the change in kinetic energy, we have

$$U + U_c = \Delta T$$

The work done by a conservative force is equal and opposite to the change in its potential energy ΔV, therefore

$$U - \Delta V = \Delta T$$

and the work–energy equation can now be written as

$$U = \Delta T + \Delta V \qquad\qquad (13.14)$$

which states that the work done by all the forces **other than the conservative forces** acting on a particle during an interval of its motion equals the sum of the corresponding changes in its kinetic and potential energy.

If the system is a conservative one, that is to say there are no forces acting on the system other than conservative forces, then $U = 0$ and

$$\Delta T + \Delta V = 0 \qquad\qquad (13.15)$$

Suppose that the kinetic and potential energy were respectively T_1 and V_1 at the beginning and T_2 and V_2 at the end of the interval. Then,

$$(T_2 - T_1) + (V_2 - V_1) = 0$$

so that

$$T_1 + V_1 = T_2 + V_2 \tag{13.16}$$

The total energy at the end of the interval equals the total energy at the beginning. In other words, in a conservative system there is no loss of mechanical energy. If the system is not conservative, as described by Equation 13.14 where perhaps work is being done by frictional forces, energy is still conserved but some mechanical energy is transformed into another form such as heat.

The principle of conservation of energy states that matter can neither be created nor destroyed but only redistributed or changed in form.

Example 13.6

A 10 kg weight is attached to one end of a light elastic spring which is connected by its other end to a horizontal bar so that it can hang free and unimpeded in the vertical plane. The weight is supported so the spring is unextended. If the weight is released, how far will it fall before it reaches its lowest position if the spring stiffness is $500\,\mathrm{N\,m^{-1}}$?

Solution   As the spring is light and elastic, it is reasonable to assume that only gravity and the action of the spring do work on the weight. In other words, all the forces are conservative and Equation 13.16 is valid. Substituting for the kinetic energy T and elastic and gravitational potential energies V_e and V_g in Equations 13.6, 13.10 and 13.12 respectively leads to

$$\frac{1}{2}mv_1^2 + \frac{1}{2}K\delta_1^2 + mgh_1 = \frac{1}{2}mv_2^2 + \frac{1}{2}K\delta_2^2 + mgh_2$$

When the weight is released, and $v_1 = 0$, it will continue to fall under gravity until it is brought to rest at its lowest point, when $v_2 = 0$; then it reverses its direction and begins to rise again under the action of the spring. If the position of the supported weight is taken as the datum, then $h_1 = \delta_1 = 0$ and $h_2 = \delta_2$. Therefore

$$\frac{1}{2}mx0 + \frac{1}{2}Kx0 + mgx0 = \frac{1}{2}mx0 + \frac{1}{2}K\delta_2^2 + mg(-\delta_2)$$

The spring extension when the weight is at its lowest point is therefore given by

$$\delta_2 = \frac{2mg}{K} = \frac{2 \times 10 \times 9.81}{500} = 0.392\,\mathrm{m}$$

When it is released, the weight will fall **0.392 m** before it reaches its lowest point.

This is the spring deflection under dynamic loading when the weight is accelerating under the action of unbalanced forces. If the weight were suspended from the spring in a state of equilibrium, so the upward force exerted by the spring exactly balanced the pull of the gravitational field, the spring would remain stationary and

$$K\delta = mg$$

so the static extension of the spring would be

$$\delta = \frac{mg}{K} = \frac{10 \times 9.81}{500} = 0.196 \text{ m}$$

which is half the dynamic deflection. In engineering applications it is very important to distinguish between static and dynamic loading conditions.

➤ **Example 13.7**

A block of mass 100 kg is released from rest at the top of a 30° incline, as shown in Figure 13.3. Calculate its speed when it has travelled 30 m down the slope (a) ignoring friction and (b) if the coefficient of kinetic friction μ_k is 0.2.

Solution   If the block moves 30 m down the slope, it descends through a vertical height of $30 \sin 30° = 15$ m. For part (a) the only work that is being done is due to the gravitational field, which is conservative, so Equation 13.15 applies. This can be written as Equation 13.16.

$$T_1 + V_{g1} = T_2 + V_{g2}$$

Taking the top of the slope as the datum ($h_1 = 0$) for the gravitational potential energy and bearing in mind that the body was released from rest ($v_1 = 0$) gives

$$\frac{1}{2} \times 100 \times 0 + 100 \times 9.81 \times 0 = \frac{1}{2} \times 100 \times v_2^2 + 100 \times 9.81 \times (-15)$$

Therefore the final velocity is given by

$$v_2 = \sqrt{\frac{100 \times 9.81 \times 15}{\frac{1}{2} \times 100}} = 17.2 \text{ m s}^{-1}$$

In part (b) the system is no longer a conservative system because work done against friction cannot be recovered. In this case we have to use Equation 13.14. The first step is to calculate the work done on the body by friction. Component at right angles to the plane give

$$mg \cos 30° - N = 0$$

The frictional force can now be found from the normal reaction N.

$$F = \mu_k N = \mu_k mg \cos 30°$$

$$= 0.2 \times 100 \times 9.81 \times \cos 30° = 169.9 \text{ N}$$

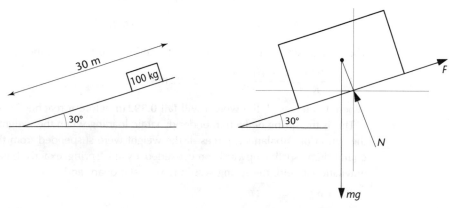

Figure 13.3 See Example 13.7.

The frictional force is resisting the motion, so it is doing negative work on the block of $U = -Fs$, where s is the distance travelled down the slope. Equation 13.14 can therefore be written

$$-Fs = (T_2 - T_1) + (V_{g2} - V_{g1})$$

$$= \left(\frac{1}{2}mv_2^2 - 0\right) + (mgh_2 - 0)$$

It was seen in part (a) that T_1 and V_{g1} are equal to zero. Rearranging the equation, we have

$$v_2 = \sqrt{\frac{mgh - Fs}{\frac{1}{2}m}}$$

$$= \sqrt{\frac{100 \times 9.81 \times 15 - 169.9 \times 30}{\frac{1}{2} \times 100}} = \mathbf{13.9\ m\,s^{-1}}$$

In part (a) the potential energy lost by the block as it descended the slope was recovered in the form of kinetic energy. In part (b) some of the potential energy has been used to overcome friction, so the kinetic energy and therefore the velocity of the block is smaller at the bottom of the slope than it was for part (a).

Recap

■ **Integrating Newton's second law with respect to displacement leads to the concepts of work and energy.**

■ **Work may be expressed in terms of the product of the displacement of a body and the component of the resultant force acting on it in the direction of the displacement.**

■ **The kinetic energy of a body is the energy it possesses by virtue of its motion and is equal to the work done in realizing that motion starting from a state of rest. This principle is embodied in the work–energy equation.**

■ **The potential energy of a body is the energy it possesses by virtue of its position and is equal to the work done in bringing it to its position from a specified datum.**

■ **Forces associated with potential energy are described as conservative forces because the work done in realizing that potential energy is fully recoverable for further work.**

■ **Systems of forces in which all the forces are conserved are known as conservative systems. A system including friction is not a conservative system because the work done in overcoming friction cannot be recovered.**

■ **The principle of conservation of energy states that matter can neither be created nor destroyed but only redistributed or changed in form. In a conservative system there is no loss in mechanical energy.**

➤ ## 13.6 Impulse and momentum

The concepts of work and energy were developed by integrating Newton's second law with respect to displacement. The concepts appear to be very useful in dealing with problems formulated in terms of forces acting over distances. For problems couched in terms of changes occurring with time, more useful are the concepts of impulse and momentum, developed by integration of Newton's second law with respect to time.

The momentum of a body, sometimes described as a measure of the quantity of its motion, has already been introduced as the product of its mass and velocity. From Newton's second law it follows that the resultant of all forces acting on a body equals its rate of change of **linear momentum** with respect to time. The momentum is described as linear to distinguish it from **angular momentum** which arises from a similar statement relating to moments, namely that the moment about a point of all the forces acting on a body equals the rate of change of angular momentum with respect to time of the body about that point. The following sections will be restricted to linear momentum.

Newton's second law has been written in terms of the momentum as Equation 10.1; with an appropriate choice of units it becomes

$$F = \frac{d}{dt}(mv) = \frac{dG}{dt}$$ (13.17)

where the symbol G represents momentum. Integrating the equation with respect to time leads to

$$\int_{t_1}^{t_2} F \, dt = \int_{G_1}^{G_2} dG$$

$$= G_2 - G_1 = \Delta G$$ (13.18)

The term on the left-hand side of the equation is known as the **linear impulse**, I. The equation states that the linear impulse on a body equals the corresponding change in linear momentum of the body. That is

$$I = \Delta G = mv_2 - mv_1$$ (13.19)

The units of impulse are generally specified as $kg\,m\,s^{-1}$, since these are the units of momentum although the units $N\,s$ are also sometimes used.

➤ ### Example 13.8

Over a period of 5 s a force of 20 N is imposed on a body with a mass of 10 kg travelling with a velocity of $5\,m\,s^{-1}$. If the force is acting in the direction of travel, calculate the impulse on the body and its new velocity.

Solution Since the force is constant, i.e. unchanging with time, the left-hand side of Equation 13.18 can be written

$$\int_{t_1}^{t_2} F \, dt = F \int_{t_1}^{t_2} dt = F(t_2 - t_1)$$

Therefore, the impulse I on the body is given by

$$I = 20(5 - 0) = 100 \text{ kg m s}^{-1}$$

From Equation 13.19 it can be seen that the final momentum mv_2 equals the sum of the impulse I and the initial momentum mv_1, so the final velocity v_2 can be found from

$$v_2 = (I + mv_1)/m = (100 + 10 \times 5)/10 = 15 \text{ m s}^{-1}$$

The impulsive force of **100 kg m s^{-1}** acting on the body increases its velocity from 5 m s^{-1} to **15 m s^{-1}**.

The applied force was constant in Example 13.8, but in some problems the applied force varies with time. The concepts of impulse and momentum are still valid but the impulse has to be treated slightly differently.

Example 13.9

If a force varying linearly with time (i.e. a graph showing how the force changes with time is represented by a straight line) and given by $F(t) = 20t$ N were applied to the body in Example 13.8, instead of the constant force of 20 N, what would be the impulse and the final velocity?

Solution   In this case the impulse is calculated as follows:

$$I = \int_{t_1}^{t_2} F \, dt = \int_0^5 20t \, dt$$

$$= 20 \left[\frac{1}{2} t^2 \right]_0^5 = 10[5^2 - 0] = 250 \text{ kg m s}^{-1}$$

This answer is the same as would have been obtained if the area under the graph of F plotted against t had been calculated between the limits of $t = 0$ and $t = 5$ s. The resulting velocity can be calculated in the same way as in Example 13.8. The time-varying force applied over an interval of 5 s gives an impulse of **250 kg m s^{-1}** and increases the velocity from 5 m s^{-1} to **30 m s^{-1}**.

Examples 13.9 and 13.10 have considered two classes of problem involving the calculation of impulse. In the first problem the applied force was constant and in the second it varied with time. A third important class considers large forces applied over very short periods of time. The term **impulsive force** is most readily associated with this class of problem, and sometimes **impact force**. Before going on to discuss this type of problem, it is helpful to generalize the concepts of impulse and momentum to systems involving more than one body.

13.7   Conservation of linear momentum

Suppose two bodies of mass m_A and m_B moving initially with velocities v_A and v_B come into contact and interact over a time interval t. If body A exerts a force F_{AB} on body B, then body B will exert an equal and opposite force F_{BA} on A, providing no other external forces are involved. Therefore

$$F_{AB} = -F_{BA}$$

And it follows from Equation 13.17 that

$$F_{AB} = \frac{d}{dt}(m_B v_B) \qquad F_{BA} = \frac{d}{dt}(m_A v_A)$$

Combining these equations leads to

$$\frac{d}{dt}(m_B v_B) = -\frac{d}{dt}(m_A v_A)$$

which can be written as

$$\frac{d}{dt}(m_B v_B + m_A v_A) = 0$$

The terms in parentheses represent the total momentum of the system, G. If the rate of change of the total momentum with time is zero, the momentum itself must remain constant, so

$$G = m_B v_B + m_A v_A = \text{constant} \qquad (13.20)$$

This equation expresses the principle of the conservation of linear momentum, which can be stated as follows. The total linear momentum of a body, or a system of bodies, in any one direction remains constant unless acted on by a resultant force in that direction.

> ### Example 13.10

A railway wagon of mass 60 Mg mass is travelling at 1.5 m s^{-1} along a horizontal track in a shunting yard when it collides with a second wagon of mass 80 Mg moving in the same direction at 1 m s^{-1}. If the wagons become coupled and move off together, calculate their common velocity and the energy loss due to the impact.

Solution   It is convenient to label the wagons A and B and their velocities before and after impact as v_{1A}, v_{1B}, v_{2A} and v_{2B} respectively. It then follows from the law of conservation of linear momentum that

$$m_A v_{1A} + m_B v_{1B} = m_A v_{2A} + m_B v_{2B} \qquad (13.21)$$

The problem cannot be solved by the momentum equation alone because it is a single equation containing two unknowns, v_{2A} and v_{2B}. However, a second independent equation involving these quantities is provided from the information that the two wagons move off with a common velocity, v_c, so that

$$v_{2A} = v_{2B} = v_c \qquad (13.22)$$

Combining Equations 13.21 and 13.22 and rearranging leads to the following expression for the common velocity of the two wagons:

$$v_c = \frac{m_A v_{1A} + m_B v_{1B}}{m_A + m_B}$$

$$= \frac{60 \times 10^3 \times 1.5 + 80 \times 10^3 \times 1.0}{60 \times 10^3 + 80 \times 10^3} = \textbf{1.214 m s}^{-1}$$

bearing in mind that the mass is expressed in kilograms, although in this instance the powers of 10 multiplying the mass cancel out.

Values of the total kinetic energy before and after the collision, T_1 and T_2, are given by

$$T_1 = \frac{1}{2} m_A v_{1A}^2 + \frac{1}{2} m_B v_{1B}^2 = 107.500 \times 10^3 \text{ J}$$

$$T_2 = \frac{1}{2} m_A v_{2A}^2 + \frac{1}{2} m_B v_{2B}^2 = 103.166 \times 10^3 \text{ J}$$

so the loss of kinetic energy due to the collision is **4.33 kJ**.

Example 13.10 demonstrates that, although momentum is conserved, there is usually an energy loss resulting from a collision. The amount of energy lost depends on the nature of the colliding bodies and how they interact. The energy loss in Example 13.10 was dictated by the fact that the bodies had a common velocity after the impact.

13.8 Types of collision

In a conservative system, as described in Section 13.5, collisions occur with no loss of energy. In this case the bodies involved are perfectly elastic and they return to their original shape after collision. Their initial energy is stored as strain energy (analogous to the energy stored in springs) at impact and is released as kinetic energy as the bodies rebound from one another. A characteristic of **elastic collisions** is that the relative velocity of the bodies before impact is equal to the relative velocity after impact, but opposite in sense.

$$v_{2A} - v_{2B} = -(v_{1A} - v_{1B}) \tag{13.23}$$

Although elastic collisions do not occur in nature, there are many problems that can be idealized as such; the collision of two hardened steel ball-bearings is just one example.

In **inelastic collisions** the bodies involved coalesce, or become mechanically linked upon impact, and their relative velocity after the collision is zero – that is they move away together, as in Example 13.10. Another example of an inelastic collision is a steel ball-bearing being dropped onto soft ground.

Partially elastic collisions lie between these two extremes. Their behaviour is governed by Newton's experimental **law of impact**, which can be summarized as follows: the ratio of the relative velocity after impact to the relative velocity before impact remains constant for a given pair of bodies, the two relative velocities having opposite sign. The constant of proportionality e is known as the **coefficient of restitution**. This is expressed by the equation

$$\frac{v_{2A} - v_{2B}}{v_{1A} - v_{1B}} = -e \tag{13.24}$$

It can be seen from Equations 13.23 and 13.24 that the coefficient of restitution e for a perfectly elastic collision takes a value of unity. Similarly, the value for an inelastic collision is zero. The coefficient varies widely between these two extremes for different bodies and, strictly speaking, depends on the impact velocity. It is often considered constant for pairs of bodies of given geometry and material, being just under 1 for hardened steel spheres and around 0.2 for lead spheres.

➤ **Example 13.11**

Two bodies of masses 10 kg and 5 kg are moving in opposite directions along the same straight line at $2\,\mathrm{m\,s^{-1}}$ and $8\,\mathrm{m\,s^{-1}}$ respectively. Calculate the energy loss on their collision if their coefficient of restitution e is (a) 1 and (b) 0.2.

Solution   (a) Assuming the bodies collide head on and assuming the 10 kg body is moving from left to right, the momentum equation gives

$$m_A v_{1A} + m_B v_{1B} = m_A v_{2A} + m_B v_{2B}$$
$$10 \times 2 + 5 \times (-8) = 10 \times v_{2A} + 5 \times v_{2B}$$

which simplifies to

$$2v_{2A} + v_{2B} = -4 \tag{13.25}$$

Applying the equation for the coefficient of restitution

$$v_{2A} - v_{2B} = -e(v_{1A} - v_{1B}) = -1[2 - (-8)] = -10 \tag{13.26}$$

Adding Equations 13.25 and 13.26 gives

$$3v_{2A} + 0 = -4 - 10 = -14$$

so v_{2A} is $-14/3\,\mathrm{m\,s^{-1}}$ and, from Equation 13.26, v_{2B} is $16/3\,\mathrm{m\,s^{-1}}$. The loss in kinetic energy due to the collision is given by

$$\Delta T = \frac{1}{2}m_A v_{1A}^2 + \frac{1}{2}m_B v_{1B}^2 - \left(\frac{1}{2}m_A v_{2A}^2 + \frac{1}{2}m_B v_{2B}^2\right)$$
$$= \frac{1}{2} \times 10 \times 2^2 + \frac{1}{2} \times 5 \times 8^2 - \left[\frac{1}{2} \times 10 \times \left(-\frac{14}{3}\right)^2 + \frac{1}{2} \times 5 \times \left(\frac{16}{3}\right)^2\right]$$
$$= 180 - 180 = \mathbf{0} \tag{13.27}$$

If the coefficient of restitution is 1, the two bodies reverse direction upon impact. This is shown by the change in the sign of their velocities. Also there is no loss in energy. This is to be expected because a coefficient of restitution of 1 represents a perfectly elastic collision for which there is no energy loss because the system is conservative.

(b) If the coefficient of restitution is 0.2, Equation 13.26 becomes

$$v_{2A} - v_{2B} = -0.2(2 + 8) = -2$$

Combining this with Equation 3.25, which remains unchanged, gives velocities v_{2A} and v_{2B} of $-2\,\mathrm{m\,s^{-1}}$ and 0 respectively. Substituting these new velocities into Equation 13.27 gives a loss in kinetic energy of **160 J**. The first body still has its motion reversed but it rebounds with a smaller velocity. Because of the energy loss on impact, there is no longer the available energy in the system to reverse the motion of the second body, which is brought to an abrupt halt by the collision.

In the examples considered up to now, the problems have been limited to the case of direct central impact in which the initial and final velocities of the bodies remain collinear. The same principles can be applied to oblique collisions by applying the law of restitution in the direction of the normal common to the initial and final trajectories, although this will not be pursued here. The final aspect of impulse and momentum to be dealt with relates to impact forces.

13.9   Impact forces

When a constant force F acts on a body for a period of time t, the impulse is Ft. If a very large force is applied over a period of time so small as to be infinitesimal, it is known as an impulsive force or an impact force. As was seen in Section 13.7, the principle of conservation of linear momentum is valid providing there are no external forces acting on the system. In the case of impact forces, the principle can be applied over the short duration of the impact, even though external forces such as friction or gravity may be present; this is because the external forces are normally much smaller than the impact force and they can be ignored. On the same basis, if a spring is subject to an impact force it may be assumed that the spring force only comes into action after the impulse has been transmitted. Examples of impact forces are explosions and concussive processes such as pile driving.

Example 13.12

A pile-driving hammer of mass 400 kg strikes a pile of mass 250 kg with a velocity of $4.875\,\text{m s}^{-1}$. If there is no rebound and the pile experiences an average resistance force from the ground of 25.877 kN, how far does the pile sink into the ground?

Solution   Since there is no rebound, the pile and the hammer move together after the impact with a common velocity v_{hp} and a combined mass m_{hp}. Although an external force acts on the pile, resisting its motion, it can be ignored for the impact over which momentum may be deemed conserved. Therefore, it follows from Equation 13.20 that

$$m_h v_{1h} + m_p v_{1p} = m_h v_{2h} + m_p v_{2p}$$
$$= (m_h + m_p) v_{hp} \tag{13.28}$$

The common velocity of the hammer and pile is given by

$$v_{hp} = \frac{m_h v_h + m_p v_p}{m_h + m_p} = \frac{400 \times 4.875 + 0}{400 + 250} = 3\,\text{m s}^{-1}$$

The work done on the pile by the ground resisting its motion may be obtained from Equation 13.14, which can be written

$$-Rs = \frac{1}{2} m_{hp} v_2^2 - \frac{1}{2} m_{hp} v_1^2 + m_{hp} g h_2 - m_{hp} g h_1$$

where the negative sign on the left-hand side of the equation indicates that the force R acts in opposition to the motion. If the pile travels a distance h through the ground, then

$$-Rh - 0 - \frac{1}{2} m_{hp} v_1^2 + m_{hp} g(-h) - 0 \tag{13.29}$$

where the initial velocity v_1 is the velocity of the pile immediately after the impact, v_{hp}. Rearranging Equation 13.29 gives

$$h = \frac{\frac{1}{2}m_{hp}v_1^2}{-R + m_{hp}}$$

$$= -\frac{-\frac{1}{2} \times 650 \times 3^2}{-25\,877 + 650 \times 9.81} = 0.150\,m$$

The pile is driven **0.150 m** into the ground by the blow from the pile-driver.

Recap

■ Integrating Newton's second law with respect to time leads to the concepts of impulse and momentum.

■ The momentum of a body, a measure of the quantity of its motion, is the product of its mass and velocity.

■ The linear impulse, obtained through integration of an applied force with respect to time over a specified interval, is equal to the corresponding change in linear momentum of a body.

■ The principle of the conservation of linear momentum states that the total linear momentum of a body, or a system of bodies, in any one direction remains constant unless acted upon by a resultant force in that direction.

■ Although momentum may be conserved in a collision, there is generally a loss of energy. A measure of the energy loss is obtained from the relative velocity of the bodies before and after impact, according to Newton's experimental law of impact.

■ The degree of energy loss varies from zero, for elastic collisions which occur in conservative systems and for which the coefficient of restitution is unity, to a maximum where the coefficient of restitution is zero and the bodies coalesce.

■ If very large forces occur over very small time intervals, as for impact forces, the principle of linear momentum can still be applied, even when external forces such as friction or gravity may be present; this is because the external forces are normally negligible compared with the impact force.

13.10 Problems

1. A locomotive of 60 tonne mass pulls some carriages of total mass 240 tonne a distance of half a kilometre up a 1 in 200 incline with an acceleration of $0.2\,\mathrm{m\,s^{-2}}$. If the overall rolling resistance is 5% of the normal reaction, what is the total work done?

2. A cyclist of mass 82 kg is riding his 13 kg bicycle up a 1 in 10 incline. What power does he need to develop in order to maintain a constant speed of 24 kph?

3. A 60 g bullet is fired at a target that presents an average resistive force to the passage of the bullet of 20 kN per metre thickness of the material from which it is constructed. If the bullet strikes the target at $600\,\mathrm{m\,s^{-1}}$, what does the minimum thickness of the target have to be to stop the passage of the bullet completely?

4. A car is proceeding down a 1 in 10 incline at 65 kph when its brakes are abruptly applied. How far does it skid along the road before stopping if the wheels lock and the coefficient of friction between the tyres and the road is 0.7?

5. A 5 kg weight whose motion is constrained by vertical guides is dropped from rest 0.5 m onto a vertically aligned spring whose stiffness is $10\,\mathrm{kN\,m^{-1}}$. How fast is the weight travelling when the spring has been compressed by 50 mm?

6. A 10 Mg truck rolling at a constant speed of 30 kph along a horizontal rail comes to an incline of slope $\sin^{-1} 0.1$. Calculate how far it will roll up the slope before coming to a halt, (a) ignoring all resistive forces, and (b) assuming the rolling resistance is 5% of the normal reaction of the truck.

7. A force of P N acting on a body of 5 kg travelling with a velocity of $5\,\mathrm{m\,s^{-1}}$ increases its velocity to $10\,\mathrm{m\,s^{-1}}$. If the force acts in the direction of motion for a period of 5 s, what is the value of P?

8. A cube of mass 10 kg is at rest on a horizontal surface when it is subjected to a horizontal force P that varies with time t as $P = 20t$. If the static and kinetic coefficients of friction are both 0.4, calculate the velocity of the cube after 5 s.

9. A 500 Mg tug is towing a 1000 Mg barge at a steady speed of $3\,\mathrm{m\,s^{-1}}$ when the towline, which is horizontal, is winched in at a rate of $0.5\,\mathrm{m\,s^{-1}}$. What is the tug's speed during this operation?

10. A railway truck of 100 kg mass travelling with a velocity of $7\,\mathrm{m\,s^{-1}}$ collides with a second truck of 200 kg mass and the two couple automatically and move off together. Calculate the velocity of the coupled trucks, and the energy lost to the system, if the second truck has a speed of $5\,\mathrm{m\,s^{-1}}$ (a) in the same direction, (b) in the opposite direction and (c) is initially stationary.

11. A body of 4 kg mass travelling at $8\,\mathrm{m\,s^{-1}}$ collides with a stationary body of 12 kg mass, both being constrained in their motion by the same frictionless guides. Find their velocities and the loss of kinetic energy, (a) if the coefficient of restitution is 1 and (b) if it is 0.5.

12. A pile of mass 50 kg is being driven into the ground by a hammer of 400 kg. If the hammer is dropped through a height of 3.5 m, there is no observable rebound, and the average resistive force of the ground on the pile is 50 kN, calculate the distance the pile sinks into the ground.

Further reading

➤ Algebra and Trigonometry

Booth, D. (1994) *Foundation mathematics*, 2nd edition. Addison Wesley Longman, Harlow.

Graham, L. and Sargent, D. (1981) *Countdown to mathematics*, book 2. Addison Wesley Longman, Harlow.

➤ Calculus

Bostock, L. and Chandler, S. (1982) *Mathematics, the core course for A-level*. Stanley Thornes, Cheltenham.

Turner, L.K. and Knighton, D. (1986) *Advanced mathematics, a unified course*, book 1. Burstall, Addison Wesley Longman, Harlow.

Answers to problems

Chapter 2

1. 22.1 km, 24.4° W of S
2. 22.1 km, 24.4° E of N
   The relationship is negative, i.e. opposite direction for the whole diagram.
3. 22.1 km, 24.4° W of S
4. 22.1 km, 24.4° E of N
   Vector P is the negative of the resultant.
5. (a) 5.00, 0; 0, −6.00; −14.1, −14.1
   (b) 5.00, 0°; 6.00, 270°; 20.0, 225°
6. 22.1 km, 24.4° W of S
7. 22.1 km, 245.6°
   Use $D_x = \sum D \cos \alpha$ and $D_y = \sum D \sin \alpha$

Chapter 3

1. 4.36P at 23.4° to the 3P force
2. $F_x = -21.2$ kN, 25.7 kN, 5.98 N 41.0 kN 0 N
   $F_y = 21.2$ kN, −30.6 kN, 0.523 N 28.7 kN 120 N
3. 59.2 kN at 37.5°
4. (a) 47.0 kN, 36.0 kN
   (b) 58.7 kN, 7.68 kN
   (c) −15.4 kN, 72.0 kN
5. (a) 19.1 N at 1.62°
   (b) 1.21 m from O
   (c) 23.1 N m
   (d) Anywhere
6. 17.3 kN at 240°
7. 86.6 N, 150 N
8. (a) 1, 1
   (b) 0.577, 1.16
   (c) 0, 1
9. (a) 0.3 kN, 1.1 kN
   (b) 0.8 kN, 2.8 kN
   (c) 0.3 kN, 1.1 kN
   (d) 1.6 kN, 0.6 kN
   (e) 0.55 kN, 1.35 kN, 2.0 kN

10. 245 N at 21.8° and 0.098 m from B
11. 6.71P at 63.4° and 0.951L from O
12. 25.8 N at 60.9°
    $Y_P = 1.12$ m

Chapter 4

1. 2.00 kN, 3.46 kN
2. 2.31 kN, 4.62 kN
3. (a) 0.5W and 0.966W at 75°
   (b) 0.577W and 1.11W at 105°
   (c) 10 kN, 5 kN, 18.5 kN at 112.5° and 10 kN at 90°
4. 48.6°, 7.2°
5. (a) 40°
   (b) 0°, W
6. 0.8P, 1.2P
7. 0.693P, 0.8P
8. 3.8W, 1.8W
9. (a) 0.433W, 0.567W
   (b) 0.433W, 0.537W
10. 7.27 kN, 16.7 kN at 296°

Chapter 5

1. 15.4 kN, −72.0 kN
2. 231 kN, 143 kN, 635 kN, 115 kN
3. 8.11 kN, 8.65 kN
4. 5.77 kN, 16.1 kN
5. 3.13 kN, 1.68 kN
6. 13.3 kN, −3.3 kN, 18.8 kN
7. 125 kN, −45 kN, 145 kN
8. 4.1 kN at 76°, 1.4 kN at 135°
9. (a) 55 kN, 45 kN, 40 kN
   (b) 11.5 kN, 51.0 kN
10. 108 kN at 141.3°, 147 kN at 63.8°
11. (a) 1.52
    (b) 13.6 kN, 71.2 kN, 24.8 kN

(d) 151.5 kN

(e) 179.3 kN

Chapter 6

1. 23.57 kN, −9.43 kN, 14.14 kN
2. 10 kN, 40.98 kN, −23.66 kN
3. 10, 10, 0; 15.80, −21.21, 10.00 (all kN)
4. −5, 5, 10; −7.90, 10.61, −3.54, −5.00 (all kN)
5. 5, 15, 10; 7.90, −10.61, −24.74, 5.00 (all kN)
6. −10.2 kN, −11.2 kN, 4.0 kN
7. −0.211 kN
8. 53.5 kN, −53.5 kN, −62.5 kN
9. 37.5 kN, 159 kN, −162.5 kN
10. 3, 14
11. 38.6 kN
12. 2, 1; 31.6, 28.3, −31.6, −23.1, 0, 0, 0 (all kN)
13. (a) 1.8 kN, 3.6 kN; 4.8 kN, 3.6 kN
    (b) 41.7 N mm^{-2}, 35.0 N mm^{-2}

Chapter 7

1. Sets (a) and (c) are right-handed.
2. (a) (r, θ, z) is $(\sqrt{2}, \pi/4, \sqrt{2})$ and (R, θ, ϕ) is $(2, \pi/4, \pi/4)$
   (b) (x, y, z) is $(\sqrt{3}, 1, 2)$ and (R, θ, ϕ) is $(2\sqrt{2}, \pi/6, \pi/4)$
   (c) (x, y, z) is $(3, 3\sqrt{3}, 2\sqrt{3})$ and (r, θ, z) is $(6, \pi/3, 2\sqrt{3})$

Chapter 8

1. $(v, t) = (1, 14.95)$, $(3, 14.65)$, $(5, 14.05)$, $(17, 4.15)$, $(19, 1.45)$.
   $(a, t) = (2, -0.15)$, $(4, -0.30)$, $(6, -0.45)$, $(14, -1.05)$, $(16, -1.20)$, $(18, -1.35)$.
2. $v = -0.0375\, t^2 + 15$, $a = -0.075t$. There is a constant difference of 0.0125 m/s for the velocities and the accelerations agree exactly.
3. $(t, v) = (0, 0.00)$, $(5, 13.81)$, $(10, 25.38)$, $(40, 61.50)$, $45, 64.31)$, $(50, 66.88)$. In 10 s the vehicle travels 132.50 m.
4. $v = t^3/3000 - t^2/20 + 2.5t$ and $s = t^4/12000 - t^3/60 + 1.25t$
5. $v = t^2 - 4t$; when $t = 0, 2$ and 4 s, $a = -4, 0$ and 4 m/s^2
6. $s = 0.25t^2 - t + 0.5$; $d = 1.25$ m.

7. $a = t/2 + 1$, $v = t^2/4 + t$, $s = t^3/12 + t^2/2$ and $v = 15$ m.

Chapter 9

1. $v_t = 75$ m/s, at $= 0$, $a_n = 11.250$ m/s^2, $a = 11,250$ m/s^2.
2. $v_t = 22.2$ m/s, $(a_l, a_n) = (-1.43, 1.796)$ m/s^2, $\ddot{\theta} = 5.19$ rad/s^2 and the number of revolutions $= 100$.
3. If v_t exceeds 19.81 m/s it begins to slide.
4. The acceleration $(a_l, a_n) = (0, 0.175)$ m/s^2
5. 30 m/s.
6. 25 m.
7. $\omega = \pi$ rad/s, maximum velocity$= 0.157$ m/s, maximum acceleration $= 0.493$ m/s^2.
8. The amplitude is 10 m and the period is 10 s.

Chapter 11

1. The plane is inclined at $11°18'$ to the horizontal.
2. The block remains stationary providing $320.5 < 2T < 660.5$.
3. 332 N.
4. 14.1 kN.
5. The force to cause sliding is 196.2 N, maximum height without toppling is 2 m.
6. The rolling resistance is 1.06% of the normal reaction.
7. 407 N.
8. 5.2 N.

Chapter 12

1. The readings are a) 9.81 m/s^2, b) 11.00 m/s^2, c) 8.62 m/s^2 and d) 11.00 m/s^2.
2. The acceleration is 0.625 m/s^2 and the tension in the tow bar is 10 kN.
3. The train covers a distance of 667 m while it is slowing down.
4. The velocity after 2 s is 11.3 m/s.
5. The tension in the bar is $4000\, \pi^2$ N.
6. The reactive force is 11.1 N directed vertically downwards.
7. The limiting coefficient of friction is 0.53.
8. a) Magnitude of force $|f| = 2.85$ kN where $(F_t, F_n) = (= -1.62, 2.34)$, b) $|f| = 3.32$ kN;

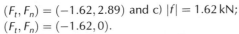

$(F_t, F_n) = (-1.62, 2.89)$ and c) $|f| = 1.62\,\text{kN}$; $(F_t, F_n) = (-1.62, 0)$.

9. The force is a) 263 N and b) −186 N.

10. The tension is 11.3 kN.

Chapter 13

1. The total work done is 44.9 MJ.
2. He develops a power of 618 W.
3. The target must be at least 0.54 m thick.
4. It skids 27.8 m before stopping.

5. 2.41 m/s.
6. a) 35.4 m b) 23.6 m
7. 5.
8. 5.38 m/s.
9. 2.67 m/s.
10. a) 5.67 m/s in the direction of the small truck, 128 J, b) 1 m/s in opposite direction; 4.8 kJ, and c) 2.33 m/s in direction of small truck; 2.1 kJ.
11. a) They rebound, each with a velocity of 8 m/s, and no loss of energy; b) they rebound, the smaller at 1 m/s the larger at 3 m/s with an energy loss of 72 J.
12. 0.268 m.

Index

**Books are to be returned on or before
the last date below.**

LIBREX—